Science and its Ways of Knowing

JOHN HATTON
University of California, Berkeley
PAUL B. PLOUFFE
University of California, Berkeley

Prentice Hall
Upper Saddle River, New Jersey 07458

Library of Congress Cataloging-in-Publication Data
Hatton, John.
Science and its ways of knowing / John Hatton, Paul B. Plouffe.
p. cm.
Includes bibliographical references.
ISBN 0–13–205576–7 (pbk.)
1. Science—Methodology. 2. Thoughts and thinking. I. Plouffe, Paul B. II. Title.
Q175.H35187 1997
502.8—dc20 96–36510
CIP

Executive Editor: *Alison Reeves*
Production Editor: *Joan Eurell*
Manufacturing Manager: *Trudy Pisciotti*
Buyer: *Ben Smith*
Interior Design and Page Composition: *Eric Hulsizer*
Cover Designer: *Wendy Alling Design*
Cover Art: *Louis Fishauf*

Simon & Schuster / A Viacom Company
Upper Saddle River, NJ 07458

Printed in the United States of America.

10 9 8 7 6 5 4 3 2 1

ISBN 0-13-205576-7

Prentice Hall International (UK) Limited, *London*
Prentice Hall of Australia Pty. Limited, *Sydney*
Prentice Hall of Canada, Inc., *Toronto*
Prentice Hall Hispanoamericana, S.A., *Mexico*
Prentice Hall of India Private Limited, *New Delhi*
Prentice Hall of Japan, Inc., *Tokyo*
Simon & Schuster Asia Pte. Limited, *Singapore*
Editora Prentice Hall do Brasil, Ltda., *Rio de Janeiro*

Contents

For Karolyn, Ian, Erin and Carolyn
and For Ev and Bill

Preface

Our intention in bringing these essays together is to invite students in the various sciences to go beyond the confines of their particular discipline and to think about science in more general terms—as a way of looking at (and thinking about) the natural world, as an attempt to understand it. Most students majoring in physics, say, or biology acquire a good basic understanding of their discipline as a body of knowledge. They are familiar with its principles, its governing theories, its vocabulary, etc. But they are far less likely to have thought much about broader and more basic issues—about, for instance, the relationship in science between fact and theory, about the *nature* of scientific theory, about the kinds of claims on truth that science makes. Such issues are left largely unexamined, though surely they are important in the education of young scientists. We believe that this collection of essays encourages a broader view, one that will give students a fuller appreciation of the grand enterprise we call science and thereby deepen their understanding of their own discipline.

Acknowledgments

We would like to thank a number of people for their advice and support. Tom Philippi offered wise counsel throughout the project; Peter Hore helped in both tangible and intangible ways. Rudy Klaver, Gene Petersen, and Ralph Rader listened to our ideas and graciously shared their knowledge. To these friends and colleagues our sincere thanks.

We would also like to thank our colleagues at Prentice Hall: Pamela Holland-Moritz, editorial assistant; Michael Schiaparelli, copyeditor; Joan Eurell, project manager; Ray Henderson, who encouraged us early in the project; and Alison Reeves, who as executive editor, helped see it through to completion.

Finally, our thanks to the following reviewers for their helpful suggestions: Joseph F. Allen, Clemson University; Dana Chatellier, University of Delaware; Alexander K. Dickison, Seminole Community College; Paul A. Draper, University of Texas at Arlington; Natalie Foster, Lehigh University; David Gavenda, University of Texas at Austin; Greg Hassold, GMI Engineering & Management Institute; John Hawley, University of Virginia; John D. McCullen, University of Arizona; Jim McGuire, Tulane University; Joseph Priest, Miami University; Stephen B. W. Roeder, San Diego State University; Robert S. Weidman, Michigan Technological University

The most remarkable discovery made by scientists is science itself.

Gerard Piel

General Introduction

Few scientists and teachers of science will disagree with Carl Sagan's assertion (in *Broca's Brain*, 1979) that "science is a way of thinking much more than it is a body of knowledge." For what we "know" in the scientific sense about the physical world—and, indeed, what we *mean* by "knowledge"—is a function of the means by which we *come* to know. Modern science began, around the late sixteenth and early seventeenth centuries, not so much with the acquisition of new knowledge about the world around us as with a new way of thinking about that world and the ways by which it can be understood. The new way of thinking made the new knowledge possible.

What are the basic elements of that thinking? Simply put, it was the recognition that an understanding—a scientific understanding—of nature must be based on observation/experiment, that ideas about the way the world works must be tested against the empirical evidence of nature itself. Before the scientific revolution, such ideas were accepted largely on the authority of the Ancients (especially Aristotle, 384–322 B.C.) and of religious dogma (derived largely from Thomas Aquinas, 1225–1274). After the revolution, the word even of such "authorities" was no longer good enough: ideas (hypotheses) would be subject to the challenge of empirical evidence. Thus, for example, Galileo (1564–1642) with his telescope confirmed Copernicus' (1473–1543) heliocentric theory of planetary motion (thereby rejecting the Ptolemaic geocentric one) and, through his (and others') experiments, refuted Aristotle's views regarding objects in motion—for instance, that larger masses fall faster than smaller ones. Ever since, science has progressed in its understanding of nature through the interaction of human reason and imagination on the one hand and nature's evidence on the other.

This interaction, of course, takes a variety of forms. Acquiring nature's evidence demands not only acute powers of observation but also a lively imagination and, increasingly, sophisticated experimental techniques. After all, nature does not often yield her secrets willingly, especially as we approach her deeper mysteries. But whatever form the interaction takes, it sooner or later gives rise to explanations constructed by the human mind. These explanations—that is, theories—are the means by which science most fully engages nature. Science makes no claims to ultimate truth, but through the explanatory power of its theories builds an ever more coherent picture of the physical world and thus extends our understanding of nature's laws.

By what means do scientific theories extend our understanding? What are their essential characteristics? First, they must be consistent: They must agree with or accommodate generally accepted scientific knowledge (though, of course, a truly revolutionary theory—such as Darwin's Theory of Evolution—will inevitably challenge some established ideas). Second, scientific theories unify; that is, they account for observations and data previously unexplained, sometimes even going beyond the boundaries of a particular discipline to show relationships hitherto unrecognized. Thus, for example, Newton's theories brought together Kepler's laws of celestial motion with Galileo's terrestrial mechanics and showed that both followed from the more general law of gravitational attraction. Or to take a more recent example, the theory of plate tectonics (continental drift) offers an explanation for such diverse phenomena as the presence far inland of fossilized marine life, the emergence of volcanoes, and the shape of continents.

Finally, scientific theories predict data or evidence not yet observed. They move from the known to the as yet unknown. In this sense, they foretell the future. Thus, Einstein's Theory of Relativity predicted that light bends as a result of gravitational forces (a prediction later confirmed in a famous experiment by Arthur Eddington), and Darwin's theory predicted the existence of "missing links" in the evolutionary development of, for example, Man—links later discovered by Louis Leakey and others. Perhaps an even more telling example of the predictive power of scientific theory is seen in Dmitri Mendeleev's ordering of the chemical elements according to their increasing atomic number. This theory, formulated in his Periodic Table, allowed Mendeleev to predict the existence of elements as yet undiscovered, a prediction soon confirmed by the discovery of Scandium, Gallium, and Germanium. Since Mendeleev's day, of course, many other predicted elements have found their appointed place in the Table. As these several examples suggest, this predictive power means not only that a theory can be confirmed but also that it can be falsified. Both confirmation and falsification constitute an advance in scientific knowledge.

Through theory, then, science drives towards an increasingly coherent picture of the universe. But as we have seen, theories are not developed in isolation, for the work of any particular scientist inevitably takes place in the context of contributions made by others. Molecular biologists today, for example, take as a given Watson and Crick's theory of the helical structure of the DNA molecule. And Watson and Crick themselves used models based on the atomic structure of molecules developed earlier by physicists. Even such a revolutionary idea as Einstein's Theory of Relativity was developed within the conceptual context of Newton's earlier work, which it partially displaced. As Einstein himself said, Newton's "great and lucid ideas will retain their unique significance for all time as the foundation of our whole modern conceptual structure in the sphere of natural philosophy."

But this "collaboration" among scientists is not simply a matter of building on the past. It involves the systematic sharing of ideas in the present. For just as science cannot be carried out in isolation, so it cannot flourish in secrecy. Indeed, openness and collaboration are built into the institution of science as part of its daily activities. Thus, for example, scientists routinely publish papers on their research, detailing their experimental procedures and results. Moreover, they must not only write these papers in such a way that their results can be reproduced, but to be published in the first place they must meet the standards of peer reviewers, which means, among other things, that their work must be judged both interesting and valuable. Then again, after publication, scientific papers are critiqued, perhaps challenged, and/or possibly used by others as a catalyst for their own work. Or consider the process of getting funding for research. First, the proposed project must be placed within the framework of established knowledge. Then a review panel must be persuaded that the project is worthwhile and that the researchers themselves are capable of carrying it off. When completed, the research must be written up and presented to the scientific community for scrutiny. In other words, from beginning to end, the process of science calls for openness and involves a kind of institutionally enforced collaboration and cooperation.

Underlying this cooperation, of course, is a consensus, a broad agreement within the scientific community regarding certain assumptions, standards, and ideals—assumptions, for example, about the nature of the universe and the means by which we can come to understand it more fully; standards regarding the nature and value of scientific knowledge; and ideals, for instance, of objectivity and intellectual integrity.

Such a consensus, however, should not be taken to imply that the scientific community has no room for individuality. On the contrary, originality and intellectual boldness are highly valued. This community, like any other, is made up of individuals who inevitably bring their own personalities, styles, strengths, limitations, interests, etc. to their work. One scientist is creative and original; another is systematic and predictable. Some work at the frontiers of their discipline (or what at any given time appear to be the frontiers), opening up new possibilities for investigation. Others work at the edges, or help fill in the blanks. Reliability and persistence are valued, but so are imagination and independence.

Yet for all its variety of styles, approaches, personalities, etc., science works as a collaborative community because that is the only way it *can* work. Scientific knowledge is expressed in its theories—theories confirmed (or at least not falsified) by nature. But because science always moves from the "known" to the unknown in building its explanation of how nature works, successful theories always represent an assimilation and/or extension of the work of others. Science, in other words, is practiced by individuals, but the nature and direction of the individual contribution is always constrained to some degree by the broader community. And because it ensures that scientific inquiry ultimately moves in the right direction, this communal constraint, exerted in the day-to-day processes of science, is fundamental to the extension of scientific knowledge.

The constraint, in short, is an essential feature of the collaboration. And perhaps the enterprise that has produced the vast structure of scientific knowledge is the grandest of all human collaborations. For science transcends race, gender, culture, and time: It is neither black nor white, male nor female, Russian nor American, and the contributions of Kepler, Galileo, and Newton are as important to the collaboration as those of Planck, Einstein, and Bohr. For like the great cathedrals of the Middle Ages, the grand conceptual structure that is science has been several hundred years in the making. Like them, too, this structure has its great central nave and its smaller side chapels. Here, stonecutters and masons work alongside builders and architects, for changes in direction and design are an inevitable part of *this* enterprise. But unlike those cathedrals, the structure of science will, in all likelihood, never be completed, for its domain is as wide as the universe, its boundaries as limitless as the human mind.

Remarkably, students in the sciences are seldom asked to think about the nature of scientific knowledge and the ways by which science "knows." And when they are, they are likely to think in terms of a prescribed and systematic "method," as though an understanding of nature (or some aspect of it) follows automatically if only the appropriate procedures are followed. Such thinking, of course, cannot help but lead to a closed and essentially rigid view of science and its practitioners. The result is a limited and therefore distorted notion regarding not only the nature of science but also its history and its grand achievements, to say nothing of the imagination and passion that so many of its practitioners have brought to it.

These essays are offered as a kind of corrective. They are not about the philosophy of science. Rather, for the most part, they present scientists speaking for themselves about their understanding of science and the attitudes and practices that inform their own—or others'—work. The geneticist and Nobel laureate Barbara McClintock remarked that nature is "much more marvelous than [the traditional notion of] the scientific method allows us to conceive." We believe that, taken together, these essays suggest something of the more open view of science and its methods that McClintock was implicitly calling for. Though hardly all-inclusive, the selection offers a fairly broad range of reference across the various sciences and should thus give the reader some sense of the history of science and its marvelous accomplishments.

A collection of essays such as this could, of course, be organized in a variety of ways. We have chosen to begin with several essays on "Method" because this seemed the best way to establish a context for what follows. Moving from the general to the particular, Part I suggests that though scientists share broad assumptions about the nature of the universe and of science, they proceed in a variety of ways in their search for knowledge—ways that frequently have as much to do with the individual scientist's style and personality as with the nature (or developmental state) of the discipline involved. To think in terms of a rigidly identifiable scientific method is not only to misconceive the nature of science but to drain it of its human element. This human—this individualized—quality is illustrated in the final three essays of Part I.

Since method drives towards the constructing of theories, Part II focuses on the nature and role of theory in scientific investigation. Many students in the sciences (to say nothing of those in other fields) believe that the business of science is truth and that, in science, truth is based on fact. But science has to do with understanding nature, not with establishing fixed truths. This understanding is expressed in the form of theories. In building up its picture—its explanation—of nature, however, science always moves from the known to the unknown (i.e., it seeks to *extend* its explanation). We may say, then, that if theory is the form in which scientific knowledge is expressed, it is also the primary instrument by which that knowledge is extended; that is, it shapes the direction of the extension. Like Part I, this Part moves from the general to the particular. After several essays on the nature and role of scientific theory, it ends with a description and explanation of the thought processes that led to one of the most famous of all theories.

But our understanding of nature does not proceed simply by means of scientific method, however understood. It frequently involves the kind of discovery that turns on less predictable (and less definable) factors such as accident and luck, as well as such personal traits as intuition, empathy, passion, openness to surprise, etc. that have to do with the personality of the individual scientist. "Discovery," as the biochemist Albert Szent-Gyorgyi said, "consists of seeing what everybody has seen and thinking what nobody has thought." But it can also involve noticing what everybody has seen but nobody has noticed. Discovery requires that we pay close attention. Part III examines a variety of contexts—and ways—in which these essentially non-rational factors have loomed large. Like the first two Parts, Part III moves from the general to the particular—from the broader scope of the first two essays to the more individualized focus of the last five.

No collection of essays could do justice to so rich and complex a subject as science and its ways of knowing. We hope this selection will encourage readers to think about science in new ways, to see it as a supremely creative achievement—perhaps the greatest of all human achievements.

Part I

On Scientific Method

Introduction

Most of us have been taught to see science as systematic, as proceeding according to a prescribed "method." Scientists, we have learned, begin their investigations with observations. On the basis of these observations, they develop a hypothesis. To test this hypothesis they extend it in the form of a prediction, which they then challenge through experiment and further observation. If the prediction is confirmed, the hypothesis survives and is subject to further investigation. The entire process, so the argument goes, is logically rigorous, objective, impersonal. Precision is of the essence. In this view, all scientific investigations proceed in much the same way, and that way is largely unaffected by personal style or the nature of the scientific discipline involved.

But scientific practice is a good deal less tidy than this view suggests. Consider, for example, the notion that scientific inquiry follows a prescribed sequence of steps starting with observation. The problem with this view is that it does not come close to describing the process by which, for example, Einstein arrived at his revolutionary notions about time and space. For Einstein started not with observational data but with a kind of intellectual puzzle (about the velocity of light as measured by different observers). Not until many years later, and well after his "solution" to the puzzle (now in the form of a theory) was generally accepted by the scientific community, were his "ideas" confirmed by observation. The history of science offers many examples of such "deviations" from the observation-hypothesis model. Or again, consider the differences in emphasis among the various sciences. Some rely heavily on theory (e.g., particle physics) while others depend primarily on observation (e.g., entomology). Moreover, variations in emphasis and focus can frequently be found even within a given discipline. Quantum chemistry, for instance, is driven principally by theory while synthetic chemistry is largely experimental. Or to take an example from the biosciences, whereas evolutionary biology is primarily concerned with an organism's change over vast periods of time, molecular biology is by and large focused on the mechanism of the cell in its present state of evolution.

But differences in the ways scientists proceed have to do with much more than the requirements or emphases of a particular discipline. For science is, above all, a human activity, and scientists, though they share certain general assumptions about the nature of the universe and the means by which we can come to understand it, inevitably bring to their work the particular talents and limitations (as well as personalities) that mark them as individuals—imagination, creativity, intuition, for instance, but also bias, shortsightedness, timidity, etc. Thus, for example, one scientist stays with a strict and safe reading of experimental results while another reaches for a bold (and risky) interpretation that offers new understanding.

Given the ways in which science actually gets done, a less rigid and more inclusive view of scientific method—one that acknowledges the complexities of scientific practice—is clearly needed. Taken together, the essays in Part I suggest such a view. In the first essay, Carl Sagan lays the groundwork by asking two fundamental questions: Can we know the universe? To what extent can we know it? The answer to these questions, he suggests, depends on both the ultimate nature of the universe itself and the intellectual capacity and openness of mind we bring to it. Openness of mind surely implies something more than rigidly logical thought. At the same time, such thought does, of course, play an important part in scientific investigation, as Robert Pirsig illustrates in the next essay. George Kneller approaches scientific method from a different perspective in the third selection. He argues that method is most usefully seen not as a predetermined sequence of steps but rather as a cycle of activities whose order is dictated by a particular investigation. For him, this cycle is similar to the process of "thoughtful problem-solving in everyday life." Emphasizing the variety of approaches among and within various sciences, Henry H. Bauer, in "The So-called Scientific Method," argues that an oversimplified view of science's methods and practices leads to a misunderstanding of the very nature of science—its grand intellectual achievement, its history, its diversity, to say nothing of the passion its practitioners frequently bring to it.

The final three essays of Part I focus on individual cases. In "The Germs of Dissent," Harry Collins and Trevor Pinch take up the matter of scientific experimentation. Focusing on the debate between Louis Pasteur and those who believed in the spontaneous generation of life, the authors show that Pasteur triumphed in part because he was willing to ignore experimental evidence (that seemed to undermine his position) simply on the basis of his conviction that spontaneous generation was impossible. In "How Fermi Would Have Fixed It," Hans Christian von Baeyer describes Enrico Fermi's distinctive and highly personal approach to science. In addition to a healthy regard for the limits of logical deduction, von Baeyer points out, Fermi had a strong sense of the value of rough-and-ready answers. Whereas most of us have been led to believe that science relies exclusively on precision and exactness, Fermi recognized that approximations are essential in scientific practice and, indeed, are no less "scientific" for being approximate. In "Sensory Function in the Harbor Seal," Deane Renouf describes her investigations into the remarkable adaptation of harbor seals to their amphibious environment. Her description suggests that scientific inquiry is a far more interactive process than is commonly believed. For although her investigation involves observation, hypothesis, experimentation, etc., its direction follows no predetermined course but rather is shaped by questions generated in the very process of the investigation. In other words, the direction of her inquiry is the product of a continuing give-and-take—a conversation, if you will—between the scientist and nature. And since this conversation is couched in terms of evolutionary theory, we might even feel the presence of a third party—Darwin himself—helping to frame the questions and thereby giving force to nature's answers.

Carl Sagan

Can We Know The Universe? Reflections on a Grain of Salt*

A native of Brooklyn, Sagan earned his B.A., B.S., M.S. and Ph.D. at the University of Chicago before he was twenty-six. He has been a professor of astronomy and astrophysics at Cornell since 1968 and is the director of its Laboratory for Planetary Studies. Much of his research has been devoted to exobiology (the possibility of extraterrestrial life), and with his associates, he has succeeded in creating amino acids from basic chemicals through the use of radiation, thus lending support to the possibility that life may exist elsewhere in the cosmos. He has played major roles in many NASA projects, including the Mariner, Viking, and Voyager missions. His numerous scientific honors include the Apollo Achievement Award given by NASA for distinguished accomplishment, while his literary awards include the 1978 Pulitzer Prize for Literature. Long an iconoclast, Sagan's provocative views on everything from nuclear disarmament to the possibility of life on other planets have made him one of modern science's most popular spokesmen.

In the following essay, taken from Broca's Brain *(1979), Sagan asks to what extent we can know the universe. Does the very nature of the universe place limitations on our knowledge? How far can common sense take us in our drive to understand? We live in a world that includes much that is knowable and much that is not. And this, he suggests, is the way it should be. This piece raises a number of issues discussed elsewhere in this collection. See for instance the essays by Kneller (Part I), von Baeyer (Part II), and Wertheimer (Part II).*

> *Nothing is rich but the inexhaustible wealth of nature. She shows us only surfaces, but she is a million fathoms deep.*
>
> RALPH WALDO EMERSON

Science is a way of thinking much more than it is a body of knowledge. Its goal is to find out how the world works, to seek what regularities there may be, to penetrate to the connections of things—from subnuclear particles, which may be the constituents of all matter, to living organisms, the human social community, and thence to the cosmos as a whole. Our intuition is by no means an infallible guide. Our perceptions may be distorted by training and prejudice or merely because of the limitations of our sense organs, which, of course, perceive directly but a small fraction of the phenomena of the world. Even so straightforward a question as whether in the absence of friction a pound of lead falls faster than a gram of fluff was answered incorrectly by Aristotle and almost everyone else before the time of Galileo. Science is based

---

*From *Broca's Brain* by Carl Sagan. ©1974, 1975, 1976, 1977, 1978, 1979 by Carl Sagan. Reprinted by permission of the author.

on experiment, on a willingness to challenge old dogma, on an openness to see the universe as it really is. Accordingly, science sometimes requires courage—at the very least the courage to question the conventional wisdom.

Beyond this the main trick of science is to *really* think of something: the shape of clouds and their occasional sharp bottom edges at the same altitude everywhere in the sky; the formation of a dewdrop on a leaf; the origin of a name or a word—Shakespeare, say, or "philanthropic"; the reason for human social customs—the incest taboo, for example; how it is that a lens in sunlight can make paper burn; how a "walking stick" got to look so much like a twig; why the Moon seems to follow us as we walk; what prevents us from digging a hole down to the center of the Earth; what the definition is of "down" on a spherical Earth; how it is possible for the body to convert yesterday's lunch into today's muscle and sinew; or how far is up—does the universe go on forever, or if it does not, is there any meaning to the question of what lies on the other side? Some of these questions are pretty easy. Others, especially the last, are mysteries to which no one even today knows the answer. They are natural questions to ask. Every culture has posed such questions in one way or another. Almost always the proposed answers are in the nature of "Just So Stories," attempted explanations divorced from experiment, or even from careful comparative observations.

But the scientific cast of mind examines the world critically as if many alternative worlds might exist, as if other things might be here which are not. Then we are forced to ask why what we see is present and not something else. Why are the Sun and the Moon and the planets spheres? Why not pyramids, or cubes, or dodecahedra? Why not irregular, jumbly shapes? Why so symmetrical, worlds? If you spend any time spinning hypotheses, checking to see whether they make sense, whether they conform to what else we know, thinking of tests you can pose to substantiate or deflate your hypotheses, you will find yourself doing science. And as you come to practice this habit of thought more and more you will get better and better at it. To penetrate into the heart of the thing—even a little thing, a blade of grass, as Walt Whitman said—is to experience a kind of exhilaration that, it may be, only human beings of all the beings on this planet can feel. We are an intelligent species and the use of our intelligence quite properly gives us pleasure. In this respect the brain is like a muscle. When we think well, we feel good. Understanding is a kind of ecstasy.

But to what extent can we *really* know the universe around us? Sometimes this question is posed by people who hope the answer will be in the negative, who are fearful of a universe in which everything might one day be known. And sometimes we hear pronouncements from scientists who confidently state that everything worth knowing will soon be known—or even is already known—and who paint pictures of a Dionysian or Polynesian age in which the zest for intellectual discovery has withered, to be replaced by a kind of subdued languor, the lotus eaters drinking fermented coconut milk or some other mild hallucinogen. In addition to maligning both the Polynesians, who were intrepid explorers (and whose brief respite in paradise is now sadly ending), as well as the inducements to intellectual discovery provided by some hallucinogens, this contention turns out to be trivially mistaken.

Let us approach a much more modest question: not whether we can know the universe or the Milky Way Galaxy or a star or a world. Can we know, ultimately and in detail, a grain of salt? Consider one microgram of table salt, a speck just barely large enough for someone with keen eyesight to make out without a microscope. In that grain of salt there are about 10^{16} sodium and chlorine atoms. This is a 1 followed by 16 zeros, 10 million billion atoms. If we wish to know a grain of salt, we must know at least the three-dimensional positions of each of these atoms. (In fact, there is much more to be known—for example, the nature of the forces between the atoms—but we are making only a modest calculation.) Now, is this number more or less than the number of things which the brain can know?

How much *can* the brain know? There are perhaps 10^{11} neurons in the brain, the circuit elements and switches that are responsible in their electrical and chemical activity for the functioning of our minds. A typical brain neuron has perhaps a thousand little wires, called dendrites, which connect it with its fellows. If, as seems likely, every bit of information in the brain corresponds to one of these connections, the total number of things knowable by the brain is no more than 10^{14}, one hundred trillion. But this number is only one percent of the number of atoms in our speck of salt.

So in this sense the universe is intractable, astonishingly immune to any human attempt at full knowledge. We cannot on this level understand a grain of salt, much less the universe.

But let us look a little more deeply at our microgram of salt. Salt happens to be a crystal in which, except for defects in the structure of the crystal lattice, the position of every sodium and chlorine atom is predetermined. If we could shrink ourselves into this crystalline world, we would see rank upon rank of atoms in an ordered array, a regularly alternating structure—sodium, chlorine, sodium, chlorine, specifying the sheet of atoms we are standing on and all the sheets above us and below us. An absolutely pure crystal of salt could have the position of every atom specified by something like 10 bits of information.* This would not strain the information-carrying capacity of the brain.

If the universe had natural laws that governed its behavior to the same degree of regularity that determines a crystal of salt, then, of course, the universe would be knowable. Even if there were many such laws, each of considerable complexity, human beings might have the capability to understand them all. Even if such knowledge exceeded the information-carrying capacity of the brain, we might store the additional information outside our bodies—in books, for example, or in computer memories—and still, in some sense, know the universe.

Human beings are, understandably, highly motivated to find regularities, natural laws. The search for rules, the only possible way to understand such a vast and complex universe, is called science. The universe forces those who live in it to understand it.

---

*Chlorine is a deadly poison gas employed on European battlefields in World War I. Sodium is a corrosive metal which burns upon contact with water. Together they make a placid and unpoisonous material, table salt. Why each of these substances has the properties it does is a subject called chemistry, which requires more than 10 bits of information to understand.

Those creatures who find everyday experience a muddled jumble of events with no predictability, no regularity, are in grave peril. The universe belongs to those who, at least to some degree, have figured it out.

It is an astonishing fact that there *are* laws of nature, rules that summarize conveniently—not just qualitatively but quantitatively—how the world works. We might imagine a universe in which there are no such laws, in which the 10^{80} elementary particles that make up a universe like our own behave with utter and uncompromising abandon. To understand such a universe we would need a brain at least as massive as the universe. It seems unlikely that such a universe could have life and intelligence, because beings and brains require some degree of internal stability and order. But even if in a much more random universe there were such beings with an intelligence much greater than our own, there could not be much knowledge, passion or joy.

Fortunately for us, we live in a universe that has, at least, important parts that are knowable. Our commonsense experience and our evolutionary history have prepared us to understand something of the workaday world. When we go into other realms, however, common sense and ordinary intuition turn out to be highly unreliable guides. It is stunning that as we go close to the speed of light our mass increases indefinitely, we shrink toward zero thickness in the direction of motion, and time for us comes as near to stopping as we would like. Many people think that this is silly, and every week or two I get a letter from someone who complains to me about it. But it is a virtually certain consequence not just of experiment but also of Albert Einstein's brilliant analysis of space and time called the Special Theory of Relativity. It does not matter that these effects seem unreasonable to us. We are not in the habit of traveling close to the speed of light. The testimony of our common sense is suspect at high velocities.

Or consider an isolated molecule composed of two atoms shaped something like a dumbbell—a molecule of salt, it might be. Such a molecule rotates about an axis through the line connecting the two atoms. But in the world of quantum mechanics, the realm of the very small, not all orientations of our dumbbell molecule are possible. It might be that the molecule could be oriented in a horizontal position, say, or in a vertical position, but not at many angles in between. Some rotational positions are forbidden. Forbidden by what? By the laws of nature. The universe is built in such a way as to limit, or quantize, rotation. We do not experience this directly in everyday life; we would find it startling as well as awkward in sitting-up exercises, to find arms outstretched from the sides or pointed up to the skies permitted but many intermediate positions forbidden. We do not live in the world of the small, on the scale of 10^{-13} centimeters, in the realm where there are twelve zeros between the decimal place and the one. Our common-sense intuitions do not count. What does count is experiment—in this case observations from the far infrared spectra of molecules. They show molecular rotation to be quantized.

The idea that the world places restrictions on what humans might do is frustrating. Why *shouldn't* we be able to have intermediate rotational positions? Why *can't* we travel faster than the speed of light? But so far as we can tell, this is the way the universe is constructed. Such prohibitions not only press us toward a little humility; they also make the world more knowable. Every restriction corresponds to a law of nature, a regular-

ization of the universe. The more restrictions there are on what matter and energy can do, the more knowledge human beings can attain. Whether in some sense the universe is ultimately knowable depends not only on how many natural laws there are that encompass widely divergent phenomena, but also on whether we have the openness and the intellectual capacity to understand such laws. Our formulations of the regularities of nature are surely dependent on how the brain is built, but also, and to a significant degree, on how the universe is built.

For myself, I like a universe that includes much that is unknown and, at the same time, much that is knowable. A universe in which everything is known would be static and dull, as boring as the heaven of some weak-minded theologians. A universe that is unknowable is no fit place for a thinking being. The ideal universe for us is one very much like the universe we inhabit. And I would guess that this is not really much of a coincidence.

Questions for Discussion

1. What does Sagan mean when he says that "science sometimes requires courage"? Can you think of a scientist who showed exceptional courage?
2. Briefly describe "the scientific cast of mind."
3. Why do the "prohibitions" of nature "make the world more knowable"?
4. Sagan says that "the testimony of our common sense is suspect at high velocities." What are some implications of this?
5. Sagan begins by asserting that "science is a way of thinking much more than it is a body of knowledge." Does the essay support this view?

Robert Pirsig

On Scientific Method*

Born in Minneapolis, the writer Robert Pirsig is best known for his autobiographical Zen and the Art of Motorcycle Maintenance: An Inquiry into Values *(1974). In this book, he uses a cross-country motorcycle trip as a framework for exploring issues ranging from the proper way to care for tools to the quandaries facing modern science. The following selection emphasizes the interweaving of inductive and deductive logic in scientific enquiry and thus particularizes issues discussed more generally by George Kneller (this Part).*

Now we follow the Yellowstone Valley right across Montana. It changes from Western sagebrush to Midwestern cornfields and back again, depending on whether it's under irrigation from the river. Sometimes we cross over bluffs that take us out of the irrigated area, but usually we stay close to the river. We pass by a marker saying something about Lewis and Clark. One of them came up this way on a side excursion from the Northwest Passage.

Nice sound. Fits the Chautauqua. We're really on a kind of Northwest Passage too. We pass through more fields and desert and the day wears on.

I want to pursue further now that same ghost that Phaedrus pursued—rationality itself, that dull, complex, classical ghost of underlying form.

This morning I talked about hierarchies of thought—the system. Now I want to talk about methods of finding one's way through these hierarchies—logic.

Two kinds of logic are used, inductive and deductive. Inductive inferences start with observations of the machine and arrive at general conclusions. For example, if the cycle goes over a bump and the engine misfires, and then goes over another bump and the engine misfires, and then goes over another bump and the engine misfires, and then goes over a long smooth stretch of road and there is no misfiring, and then goes over a fourth bump and the engine misfires again, one can logically conclude that the misfiring is caused by the bumps. That is induction: reasoning from particular experiences to general truths.

Deductive inferences do the reverse. They start with general knowledge and predict a specific observation. For example, if, from reading the hierarchy of facts about the machine, the mechanic knows the horn of the cycle is powered exclusively by electricity from the battery, then he can logically infer that if the battery is dead the horn will not work. That is deduction.

Solution of problems too complicated for common sense to solve is achieved by long strings of mixed inductive and deductive inferences that weave back and forth between the observed machine and the mental hierarchy of the machine found in the manuals. The correct program for this interweaving is formalized as scientific method.

Actually I've never seen a cycle-maintenance problem complex enough really to require full-scale formal scientific method. Repair problems are not that hard. When I think of formal scientific method an image sometimes comes to mind of an enormous juggernaut, a huge bulldozer—slow, tedious, lumbering, laborious, but invincible. It takes twice as long, five times as long, maybe a dozen times as long as informal mechanic's techniques, but you know in the end you're going to *get* it. There's no fault isolation problem in motorcycle maintenance that can stand up to it. When you've hit a really tough one, tried everything, racked your brain and nothing works, and you know that this time Nature has really decided to be difficult, you say, "Okay, Nature, that's the end of the *nice* guy," and you crank up the formal scientific method.

For this you keep a lab notebook. Everything gets written down, formally, so that you know at all times where you are, where you've been, where you're going and where you want to get. In scientific work and electronics technology this is necessary because otherwise the problems get so complex you get lost in them and confused and

forget what you know and what you don't know and have to give up. In cycle maintenance things are not that involved, but when confusion starts it's a good idea to hold it down by making everything formal and exact. Sometimes just the act of writing down the problems straightens out your head as to what they really are.

The logical statements entered into the notebook are broken down into six categories: (1) statement of the problem, (2) hypotheses as to the cause of the problem, (3) experiments designed to test each hypothesis, (4) predicted results of the experiments, (5) observed results of the experiments and (6) conclusions from the results of the experiments. This is not different from the formal arrangement of many college and high-school lab notebooks but the purpose here is no longer just busywork. The purpose now is precise guidance of thoughts that will fail if they are not accurate.

The real purpose of scientific method is to make sure Nature hasn't misled you into thinking you know something you don't actually know. There's not a mechanic or scientist or technician alive who hasn't suffered from that one so much that he's not instinctively on guard. That's the main reason why so much scientific and mechanical information sounds so dull and so cautious. If you get careless or go romanticizing scientific information, giving it a flourish here and there, Nature will soon make a complete fool out of you. It does it often enough anyway even when you don't give it opportunities. One must be extremely careful and rigidly logical when dealing with Nature: one logical slip and an entire scientific edifice comes tumbling down. One false deduction about the machine and you can get hung up indefinitely.

In Part One of formal scientific method, which is the statement of the problem, the main skill is in stating absolutely no more than you are positive you know. It is much better to enter a statement "Solve Problem: Why doesn't cycle work?" which sounds dumb but is correct, than it is to enter a statement "Solve Problem: What is wrong with the electrical system?" when you don't absolutely *know* the trouble is *in* the electrical system. What you should state is "Solve Problem: What is wrong with cycle?" and *then* state as the first entry of Part Two: "Hypothesis Number One: The trouble is in the electrical system." You think of as many hypotheses as you can, then you design experiments to test them to see which are true and which are false.

This careful approach to the beginning questions keeps you from taking a major wrong turn which might cause you weeks of extra work or can even hang you up completely. Scientific questions often have a surface appearance of dumbness for this reason. They are asked in order to prevent dumb mistakes later on.

Part Three, that part of formal scientific method called experimentation, is sometimes thought of by romantics as all of science itself because that's the only part with much visual surface. They see lots of test tubes and bizarre equipment and people running around making discoveries. They do not see the experiment as part of a larger intellectual process and so they often confuse experiments with demonstrations, which look the same. A man conducting a gee-whiz science show with fifty thousand dollars' worth of Frankenstein equipment is not doing anything scientific if he knows beforehand what the results of his efforts are going to be. A motorcycle mechanic, on the other hand, who honks the horn to see if the battery works is informally conducting a true scientific experiment. He is testing a hypothesis by putting the question to nature. The TV scientist who mutters sadly, "The experiment is a failure; we have failed to achieve what we had hoped for," is suffering mainly from a bad scriptwriter. An experiment is never a failure solely because it fails to achieve predicted results. An experiment is a failure only when it also fails adequately to test the hypothesis in question, when the data it produces don't prove anything one way or another.

Skill at this point consists of using experiments that test only the hypothesis in question, nothing less, nothing more. If the horn honks, and the mechanic concludes that the whole electrical system is working, he is in deep trouble. He has reached an illogical conclusion. The honking horn only tells him that the battery and horn are working. To design an experiment properly he has to think very rigidly in terms of what directly causes what. This you know from the hierarchy. The horn doesn't make the cycle go. Neither does the battery, except in a very indirect way. The point at which the electrical system *directly* causes the engine to fire is at the spark plugs, and if you don't test here, at the output of the electrical system, you will never really know whether the failure is electrical or not.

To test properly the mechanic removes the plug and lays it against the engine so that the base around the plug is electrically grounded, kicks the starter lever and watches the spark-plug gap for a blue spark. If there isn't any he can conclude one of two things: (a) there is an electrical failure or (b) his experiment is sloppy. If he is experienced he will try it a few more times, checking connections, trying every way he can think of to get that plug to fire. Then, if he can't get it to fire, he finally concludes that *a* is correct, there's an electrical failure, and the experiment is over. He has proved that his hypothesis is correct.

In the final category, conclusions, skill comes in stating no more than the experiment has proved. It hasn't proved that when he fixes the electrical system the motorcycle will start. There may be other things wrong. But he does know that the motorcycle isn't going to run until the electrical system is working and he sets up the next formal question: "Solve problem: what is wrong with the electrical system?"

He then sets up hypotheses for these and tests them. By asking the right questions and choosing the right tests and drawing the right conclusions the mechanic works his way down the echelons of the motorcycle hierarchy until he has found the exact specific cause or causes of the engine failure, and then he changes them so that they no longer cause the failure.

An untrained observer will see only physical labor and often get the idea that physical labor is mainly what the mechanic does. Actually the physical labor is the smallest and easiest part of what the mechanic does. By far the greatest part of his work is careful observation and precise thinking. That is why mechanics sometimes seem so taciturn and withdrawn when performing tests. They don't like it when you talk to them because they are concentrating on mental images, hierarchies, and not really looking at you or the physical motorcycle at all. They are using the experiment as part of a program to expand their hierarchy of knowledge of the faulty motorcycle and compare it to the correct hierarchy in their mind. They are looking at underlying form.

QUESTIONS FOR DISCUSSION

1. What, according to Pirsig, constitutes an experimental failure? Could an experiment that fails to achieve predicted results ever be considered a success?
2. In Pirsig's opinion what constitutes the "larger intellectual process" of which scientific experimentation is only a part?
3. Pirsig says, "When I think of formal scientific method an image sometimes comes to mind of an enormous juggernaut, a huge bulldozer—slow, tedious, lumbering, laborious, but invincible." Later in the same passage, he continues, "When you've hit a really tough one . . . and you know that this time Nature has really decided to be difficult, you say, 'Okay, Nature, that's the end of the *nice* guy,' and you crank up the formal scientific method." What view of nature—and the scientist's relationship to it—do these statements imply?

GEORGE F. KNELLER

A Method of Enquiry*

Educated at the University of London (M.A.) and Yale University (Ph.D.), George Kneller was for many years on the faculty of the University of California at Los Angeles. He has published extensively on the subject of education, especially science education, and is a Fellow of the Society for the Advancement of Science and of the National Council of Learned Societies. Among his many awards are honorary doctorates from the universities of Sheffield and Heidelberg.

In the following essay, taken from his book Science as a Human Endeavor *(1978), Kneller examines the essential character of scientific method through a discussion of the ways in which it proceeds. For him, this method is best seen not as a prescribed sequence of steps but as a cycle of activities whose order is determined by the particular problem under investigation. This more open (and pragmatic) view of method is echoed in Bauer and illustrated in Renouf's investigation of harbor seals (both in this Part).*

To the lay visitor a scientific laboratory hums with efficiency. Not so to most scientists. They know how much of research is trial and error—how much depends on factors other than scientific laws and method. As biologist J. Z. Young says:

> in his laboratory he [the scientist] does not spend much of his time thinking about scientific laws at all. He is busy with other things, trying to get some piece of apparatus to work, finding a way of measuring something more exactly, or making a dissection that will show the parts of an animal or plant more clearly. You may feel that

*Originally appeared in *Science as a Human Endeavor* by George F. Kneller. ©1978 by Columbia University Press. Reprinted with permission of the publisher.

> he hardly knows himself what law he is trying to prove. He is continually observing, but his work is a feeling out into the dark, as it were. When pressed to say what he is doing he may present a picture of uncertainty or doubt, even of actual confusion.

Is there a method in this confusion?

There *may* be. Some writers have claimed that all research projects involve the same core activities. But this claim is surely false. Hypothesizing is the essence of theory-construction, yet in ordinary fact-finding no hypotheses are invented. (A hypothesis is a research-guiding conjecture.) Hence, there is no single scientific method, in the sense of a single sequence of research acts exemplified in all varieties of research. Nevertheless, all original research projects, all investigations in which a hypothesis is formed, do involve a common cycle of activities. This cycle is strikingly like the structure of thoughtful problem-solving in everyday life. Consider an example.

In a certain city a new road had been built and the accident rate soared. There was a public outcry and an investigation followed. The investigators began with the most obvious hypothesis—that a new road increases traffic, which increases accidents. But they found that accidents had mounted disproportionately. They then conjectured that on a new road drivers were less careful. But the statistics on other new roads disproved this. So they surmised that the cause was speeding. According to police records, however, fewer drivers had been cited than usual. Had the police been less active? No, the same number had been on duty. Then the investigators noticed that most of the accidents had occurred in only three places on the road. So they recommended new rules for traffic at these points. After this the number of accidents fell below the norm. The problem had been solved.

The process I have described looks tidy, but that is because it is the structure and not the experience of an investigation. This structure often is concealed from the investigator by the acts in which it is embodied—acts that may be inconsistent or thwarted. Take another example.

Both in literature and in life, detectives eat, drink, and sleep hypotheses. They examine the scene of a crime, interview witnesses and suspects, postulate motives, propose scenarios, and test everything against their data. Some detectives can finger a culprit before the evidence is in. It comes to them "naturally," some people say, but perhaps more from experience with similar cases.

A detective's job can be quite as difficult as a scientist's. After working a string of hypotheses, the detective may get his man. It may take him a few days, a few months, a few years. Or he may fail. Most crimes are never solved. Now listen to a scientist: "Nearly all scientific research," says biologist Peter Medawar, "leads nowhere—or, if it does lead somewhere, then not in the direction it started off with. . . . I reckon that for all the use it has been to science about four-fifths of my time has been wasted, and I believe this to be the common lot of people who are not merely playing follow-my-leader in research."

Now of course the detective analogy, like all analogies, breaks down in certain respects. Detectives and scientists have different goals. The detective aims to catch a criminal; the scientist, to contribute to knowledge. Also the techniques of their investigation differ, owing to the kinds of evidence each seeks to obtain. Nevertheless, in both examples we find the same sequence of activities as in scientific research—problem, hypothesis, inference, test, feedback, change of hypothesis, and the sequence repeated. Thus the scientific method is not unique. Scientific inquiry uses more sophisticated knowledge and more refined techniques than those of everyday thoughtful problem-solving, but its rational structure is the same. Let me outline this structure more formally.

While carrying out observations or experiments, or reflecting on current knowledge, the scientist notices something unusual, such as a fact at odds with an established theory or an inconsistency among theories. (Darwin, for instance, noticed 13 species of finches on the Galapagos Islands; Einstein saw that Newtonian mechanics and Maxwell's electrodynamics were inconsistent.) He formulates the disagreement as a problem to be investigated. After more observation or reflection he proposes a solution—a hypothesis that something is the case. He then deduces the implications of this hypothesis, predicting what states of affairs should obtain if the hypothesis is correct. If these states of affairs are observable (i.e., if there are instruments that can detect them), he carries out observations or experiments to collect data on them. He compares the data with his predictions, and if the two sets of statements agree, he considers the hypothesis to that extent confirmed. If they disagree, he has three options: to make further predictions and conduct further tests; to propose another hypothesis, deduce its implications, and test them (a process he may repeat several times); or to abandon the project altogether. If his predictions are confirmed (or if he expects them to be), he writes a draft of his solution, stating his hypothesis, data, and predictions. This is the first research cycle, the cycle of *discovery*. It accommodates all the unforeseen events mentioned above, and it forms the structure of all research projects in which a hypothesis is invented.

It is followed by another, the cycle of *validation*. The scientist now submits his solution to the judgment of his peers. He therefore must relate the solution to established knowledge and show that his arguments and techniques meet the standards of the field. As a rule he presents a preliminary report at a meeting of his fellow specialists and defends it against criticism. Next he writes a formal paper which he sends to a professional journal. His solution then is checked by other scientists for soundness of reasoning, accuracy of calculation, adequacy of evidence to conclusion, and significance of the problem itself. If the solution survives repeated tests, it is accepted as reliable and used in the investigation of further problems.

Types of Research

The method I have described is used in some scientific research, but not in all. The main types of research are fact-finding, consolidation, extension, reformulation, and theory-creation.

At least half of scientific research consists of fact-finding or gathering data about phenomena already partly known, such as the positions of stars, the specific gravities of materials, wave lengths, electrical conductivities, the boiling points of solutions, and so on. It includes the testing of laws, theories, and hypotheses, and experimenting with new instruments and techniques to see what they will achieve. In such research, hypotheses ordinarily are not invented, so the method I have described is not normally used.

Consolidation consists in developing the implications of a law or theory for the areas in which it is expected to apply. During the eighteenth and early nineteenth centuries, for example, many scientists sought to predict the motions of the moon and planets from Newton's laws of motion and his law of gravitation. In 1846 Antoine Leverrier correctly predicted the existence of the planet Neptune.

Extension is the application of a law or theory to new areas. In the eighteenth century, scientists applied Newton's laws of motion to hydrodynamics and vibrating strings, and in 1905 Einstein used Planck's quantum theory to propose that light travels in photons. For half a century, dedicated physicists have sought to consolidate and extend general relativity.

Reformulation is the revision of a theory to make it clearer, simpler, or more easily applicable. During the eighteenth and nineteenth centuries, a succession of brilliant mathematicians (Euler, Lagrange, Gauss, Hamilton) reformulated Newton's laws and worked out techniques to apply them more widely and precisely. In this century scientists have worked on the mathematical and philosophic foundations of quantum mechanics.

Theory-construction, including the creation of new laws and taxonomies, is the most vital and original form of scientific research. All forms, however, other than fact-finding, entail the invention of hypotheses and hence the use of the method I have described.

Techniques

As a rule this method is used together with special techniques, acquired largely through practice during the scientist's apprenticeship. These techniques may be conceptual (such as algorithms—step by step procedures—for deducing consequences and checking solutions) or empirical (such as procedures for making observations and performing experiments). Each science has its own techniques. Biologists, for example, but not astronomers, use control groups. A scientist can divide a group of rabbits with similar characteristics, treat both sets alike on every characteristic but one, and observe the results. But he cannot do so with stars or galaxies. In chemistry some widely used techniques are solution, filtration, evaporation, distillation, and crystallization. Different branches and specialties also have characteristic techniques. Most organic chemists use spectrometers, whereas physical chemists have only the computer in common. Again, within physical chemistry many specialties are distinguished by the use of a particular instrument: flash photolysis, laser X-ray, photoelectron spectrometry, low-energy electron diffraction, and so on.

All sciences, however, use concrete models. For centuries astronomers have used the orrery, a model of the bodies and movements of the solar system. Today biochemists and molecular biologists employ models of the atomic structure of molecules. Watson and Crick made several such models on the way to their theory of the helical structure of the DNA molecule.

Facts and Data

The scientist observes facts and records them in data. *Facts* are things that happen or subsist; they are events or states. *Data* are symbolic representations of events and states—generally statements recording them. Surprising as it may seem, there is no fact that is not colored by our preconceptions. This can be shown from everyday experience, for what we mostly perceive are objects and processes of definite kinds, not raw sense impressions in "blooming, buzzing confusion." We interpret sense impressions by means of concepts, and so have perceptions rather than meaningless sensations. Thus perception is essentially interpretive or judgmental. As a contemporary philosopher has said:

> Perception must . . . be understood as the activity of referring a present sense-content to the systematically structured background knowledge of the world; and the successful outcome of this activity is the achievement of recognition. . . . But the fact has not been *given* to us gratis. We have achieved it through a complex activity of schematizing, organizing, reference, and interpretation of the scrutinized contents of primitive sentience.

In science the conceptual schemes that enter into our observations are more theoretical, more exact, and more consciously criticized than those of ordinary life. Hence both facts and data are "theory-laden." In Russell Hanson's words, "The observations and the experiments are infused with the concepts; they are loaded with the theories." This happens because the theories themselves define the concepts in terms of which the data are expressed and the facts are interpreted. As things are conceived, so are they seen. The theories not only direct our attention to things we had not noticed before, they also influence what we see when we see it. In some cases this is obvious. To a layman a photograph of a bubble chamber is a pretty picture. Only a particle physicist can read the lines and spirals as the tracks and collisions of specific particles. But it also is true of the direct observation of gross physical objects like the sun. As Hanson shows, Tycho Brahe, holding the geocentric theory, and Johann Kepler, holding the heliocentric, saw the same sun differently. Brahe saw the sun rise over the earth, but Kepler saw the earth's horizon fall away from the sun. Or take Kuhn's example of a stone swinging from a fixed support. For an Aristotelian what counted was the coming to rest of the stone in its lowest position. For Galileo what mattered was that the stone repeated its motion and instead of remaining at its lowest point swung to the opposite extreme. Thus, where the Aristotelian saw a stone prevented from failing, Galileo saw a pendulum.

Nevertheless, although the facts are theory-laden, they need not all be loaded by the theory or theories under test. At sunrise Brahe and Kepler saw the sun's orb and the horizon move apart. This fact, laden with a much older theory, was common to them both. Similarly, both Galileo and Aristotle saw a stone swinging. Since rival scientists see at least some facts alike, competing theories can be compared. Moreover, theory-loading does not make theories self-confirming. Theories determine what the facts will be like but not what they will be—what could confirm them, but not what does.

Observation

The scientist observes much more carefully than the ordinary man. The good scientist looks for the unexpected. Of Charles Darwin, his son wrote: "There was one quality of mind which seemed to be of special and extreme advantage in leading him to make discoveries. It was the power of never letting exceptions pass unnoticed." The same might have seen said of Pasteur. One day, while watching bacteria in a tiny drop of fluid undergoing butyric acid fermentation, he was astonished to see that as the organisms approached the edge of the drop they stopped moving. He guessed that this was because near the air the fluid contained oxygen. What did this fact signify? That these bacteria lived where there was no oxygen. From this insight he leapt to the conclusion that life can exist without oxygen, a condition generally thought to be impossible. This important hypothesis sprang from the observation of an anomaly that few would have noticed.

Pasteur was using a microscope. Instruments enormously increase the range and accuracy of observation. With some instruments, such as the microscope and telescope, we observe phenomena directly. With others what we observe directly is treated as evidence of something unobserved. Take the Fermi Laboratory accelerator near Chicago. This huge circular machine, 4 miles in circumference, accelerates particles to velocities near the speed of light and then slams them together. As the particles collide, they disintegrate into their constituent parts, or at any rate into some of them. The collisions may be photographed in a bubble chamber, a sphere filled with liquid hydrogen. When a beam of particles hurtles into the chamber, a piston moves, releasing the pressure. As though a bottle of champagne had been opened, bubbles form in the liquid hydrogen along the ionized paths of the particles. The tracks revealed by the bubbles are photographed, telling scientists of a world that cannot be observed directly.

Scientific observation is systematic, detailed, and varied. It is made *systematic* by being controlled by a hypothesis or by a precise idea of the phenomenon to be located. It is made *detailed* by using powerful instruments and by concentrating on particular properties of a phenomenon. It is made *varied* by observing the phenomenon under different conditions, or, in an experiment, by varying and holding constant different variables in order to note the results.

The data obtained by observation are expected to be objective, reliable, and precise. Data are *objective*, or intersubjectively testable, in the sense that any suitably trained scientist, performing the same operations, is expected to observe the same facts as those recorded and thus obtain the same data. To this end, data are expressed in the language of physical things (rocks, plants, colors, sounds, weights, pointer readings) rather than in terms of sensations unique to the individual. The data are *reliable* when the facts are given a description that different scientists, observing the facts, can accept. The data also are expected to be *precise*; they should describe a fact so as to differentiate it as much as possible from similar facts. The most objective, reliable, and precise data are quantitative.

The Research Problem

The research cycle begins not with observation or measurement, but with the search for, or location of, a problem. This problem may be empirical, such as the existence of an anomaly to a well-confirmed law or theory. In 1933, for example, Carl D. Anderson in Pasadena found evidence of what looked like a positive electron. Hitherto scientists had recognized only negatively charged electrons and positively charged protons. Yet tracks in the cloud chamber now suggested that a particle existed with the mass of an electron and a double (positive and negative) curvature. This was doubly anomalous. Positive curvature normally implied a mass 1,000 times greater than that of the electron. Moreover, a particle of such mass would have a range of 5 millimeters, whereas the track itself was 5 centimeters long. So Anderson posed this problem: Is a positive electron possible? Is this the strange particle predicted by Paul Dirac? He answered, rightly, that it was.

Or the problem may be conceptual. The Copernican theory made a number of assumptions about the motion of bodies that clashed with the established Aristotelian dynamics. One of the strongest arguments against the theory was that it was unsupported by any theory of motion that would justify Copernicus's assumptions about the motion of the earth (e.g., that the earth rotated on its axis once every 24 hours). Recognizing this conflict between the two schemes, Galileo created a new dynamics of relative motion that was compatible with the Copernican theory.

Often, however, the scientist begins research by looking for a problem. He may choose an area that is well developed theoretically and hence full of leads for further investigation; or he may go where there has been a sudden rise in the rate of empirical discovery; or he may simply have a hunch that an area is rich in possibilities. This seems to have been why physicist Bruno Rossi of MIT joined the new field of X-ray astronomy, where he discovered (1962) the first X-ray source outside the solar system, Scorpio X-1. He describes his motivation as follows:

> The initial motivation of the experiment which led to this discovery was a subconscious feeling for the inexhaustible wealth of nature, a wealth that goes far beyond the imagination of man. . . . More likely it was inborn and was the reason why, as a young man, I went into the field of cosmic rays. In any case, whenever technical progress [in this case, space science techniques] opened a new window into the surrounding world, I felt the urge to look through this window, hoping to see something unexpected.

The Hypothesis

Having formulated his problem, the scientist looks for a hypothesis. A hypothesis is a conjecture that something is the case. It normally is expressed in a statement or set of statements from which conclusions can be drawn about what else would be the case if this is. It often takes the form, "If *A* is true then *B* might (should, will) follow."

What are characteristics of a good hypothesis? First, it should account for the known facts. (Nevertheless this qualification sometimes may be disregarded, since a scientist proposing a revolutionary new theory may have to ignore some of the accepted facts while looking for new facts of his own.) Second, it must be precise enough to yield testable predictions. As such it is valuable even when incorrect, for if it can be decisively refuted, it can be eliminated as a possible solution. Third, it should predict some fact or facts hitherto unknown. Einstein, for instance, deduced three predictions from his theory of general relativity: the deflection of light in the sun's gravitational field, the motion of Mercury's perihelion, and the red shift of light from distant stars. The first two predictions were confirmed in due course, and the third agreed with existing data. But many scientists did not consider the third prediction novel, since the disagreement with Newton had been known for almost a century. However, it has recently been argued that a new fact is best seen simply as one more fact unexplained by theory. By this criterion, Mercury's perihelion and the Michelson-Morley experiment were novel facts for the general and special theories of relativity, respectively.

From Problem to Hypothesis

Sometimes a scientist leaps to a hypothesis almost as soon as he sights a problem. In 1895 Wilhelm Röntgen noticed a greenish glow coming from a cathode-ray tube in his laboratory. Thinking that the glow might be caused by ultraviolet rays, he put a fluorescent screen near the tube. It lit up. He then put the tube in a cardboard box. Still the screen lit up. This showed that the radiation could not be ultraviolet light, which does not penetrate cardboard. Röntgen reflected: the rays passed through the glass tube, the cardboard box, and the air to illuminate the fluorescent screen; they therefore must be an unknown form of invisible light, and if so, they must cast a shadow. On an impulse, Röntgen put his hand in front of the screen. To his amazement he saw not the shadow but the skeleton of his hand, the flesh and skin forming a faint, grayish fringe. He realized immediately that he was dealing with an entirely new kind of radiation. After carrying out further experiments, he published a paper describing the properties of these "X-rays," as he called them. And "X-rays" they have remained. Poor Röntgen—to have had the measuring units named in his honor, but not the rays!

A complex, revolutionary hypothesis, on the other hand, may take some time to form. During his five years on H.M.S. *Beagle* (1831–36), Darwin, as ship's naturalist, amassed a store of evidence about the plants, animals, and geological strata of South America. But he was interested mainly in geology. His visit to the Galapagos Islands, where he saw small variations in the species of birds and tortoises from island to island, shook his belief that species were immutable. In 1835 he proposed a theory of coral reefs that resembled his later theory of evolution. But he did not become an evolutionist until nearly two years after his return to England. At this point he put forward two different evolutionary hypotheses, only to reject them. Then he began to search for the specific cause or causes of evolution. The idea that natural selection is the cause struck him over a year later while he was reading Malthus. Darwin described this insight in his *Autobiography*:

> In October 1838, that is fifteen months after I had begun my systematic enquiry, I happened to read for amusement Malthus on *Population*, and being well prepared to appreciate the struggle for existence that everywhere goes on from long-continued observation of the habits of animals and plants, it at once struck me that under these circumstances favorable variations would tend to be preserved and unfavorable ones to be destroyed. The result of this would be the formation of new species. Here, then, I had at last got a theory by which to work.

Darwin did not write a draft of his hypothesis until four years later, when he produced an abstract of 35 pages. Two years afterward he enlarged this abstract to 230 pages. For the next 14 years he discussed the theory with correspondents but did not announce it publicly. Then he received a paper from Alfred Russel Wallace giving a brief statement of the theory, which Wallace had thought up independently. A joint paper by Darwin and Wallace was read before the Linnaean Society of London on July 1, 1858, and then published. Darwin's book, *The Origin of Species*, appeared on November 24, 1859, and immediately booksellers bought up the entire edition of 1,500 copies.

Why did Darwin wait so long to announce his theory? Partly because he wished to perfect the theory, but mainly because he feared persecution for advocating a view that might be stigmatized as a rejection of the Biblical account of Creation. As Darwin knew, Galileo, at 70, had been forced by the Inquisition to renounce the Copernican theory, and Giordano Bruno, an earlier supporter of that theory, had been burned at the stake. Although he did not expect so grim a fate, Darwin nonetheless hesitated to defy public opinion.

After proposing a firm hypothesis, what does the scientist do with it? As a rule he tests the hypothesis by making observations or carrying out an experiment. The test often is inconclusive. So, with the aid of data provided by the test, the scientist may refine his hypothesis through a series of trials (i.e., reformulations and tests), each trial furnishing data with which he can make the hypothesis more precise. Alternatively, instead of refining a single hypothesis, he may test a series of different ones. Trying to calculate the orbit of the planet Mars, Kepler began with the hypothesis that it was circular. He found this hypothesis refuted by the facts, but tried it again, only to find it refuted more decisively still. So he added a bulge to one side of the circle and made it an ovoid. Since this hypothesis turned out to be self-contradictory, he modified the ovoid so that it began to resemble an ellipse. Finally, he proposed that the form of a planet's orbit is a perfect ellipse.

Another tactic is to propose a number of hypotheses at the outset and eliminate them successively until only one remains. Alternatively, the scientist may fit them one by one into a general scheme. Thus Darwin took a first hypothesis from domestic breeding, a second from the struggle for existence and the natural selection of wild varieties, a third from the ramification of species from common progenitors as shown in the geological record, and a fourth from the geographical distribution of species. He combined these hypotheses to form his theory of the evolution of species.

In the course of a research project the scientist gathers data as well as invents hypotheses. After most tests he uses the data to modify or replace his hypothesis. Let us see how Lord Rayleigh, for example, proceeded through a succession of hypotheses and tests to the discovery of the gas argon. In 1892 Rayleigh discovered that atmospheric nitrogen is one-half percent heavier than nitrogen prepared from chemical compounds. Why was this? His first hypothesis was that when nitrogen is chemically prepared, it becomes mixed with a light gas such as hydrogen. As a test he introduced hydrogen into nitrogen, but found the density unaffected. Out went that hypothesis. The alternative was that there is an unknown heavy gas in the atmosphere. The way to test this was to eliminate the real nitrogen from the atmosphere and see if anything was left. So he mixed nitrogen with oxygen and sent electric sparks through the mixture. (Nitrogen and oxygen combine, resulting in a compound that can be removed.) The experiment dragged on for almost two weeks. The sparking apparatus kept stopping, and Rayleigh would doze in an armchair in an adjoining room until late at night with a telephone near him to relay the noise of the instrument. When the noise stopped, he woke and adjusted the apparatus. Eventually a small residue was left. Was this hydrogen or nitrogen? He performed a couple of experiments to disprove the first possibility. First he passed atmospheric nitrogen over red-hot magnesium, leaving a small residue, which turned out to be heavier than nitrogen and thus heavier than hydrogen. Then he carried out an atmolysis of air—a process in which a mixture of gases is leaked through a porous pot—and found a small residue, again heavier than nitrogen. So he proposed that there exists a hitherto unknown gas, argon. He confirmed this hypothesis by carrying out a number of tests eliminating the possibility that the gas was hydrogen. Argon also was found to have a different spectrum than nitrogen and to be two and a half times as soluble in water. This last finding suggested that there should be more argon than nitrogen in rainwater, which tests again confirmed.

Some Forms of Reasoning

I now want to consider several modes of reasoning leading to the formation of hypotheses that have been described by philosophers of science. These are retroduction, hypothetico-deduction, induction, and reasoning by analogy.

In the case of *retroduction* (R-D), the scientist encounters an anomaly and then seeks a hypothesis from which the existence of the anomaly can be deduced. Thus he reasons back from the anomaly to a hypothesis that will explain it. The form of the inference is this: An anomalous fact *A* is observed; *A* would be explicable if hypothesis *H* were true; hence, there is reason to think that *H* is true. Kepler, for instance, reasoned retroductively to his hypotheses about the orbit of Mars. He began by proposing that Mars moves in a perfect circle. But he found that the predictions he deduced from this hypothesis conflicted with the data of the Danish astronomer Tycho Brahe. The data, therefore, appeared to be inconsistent with the hypothesis of circular motion. So he assumed that the data were correct and sought to explain them by proposing the hypotheses we have mentioned.

Instead of reasoning to a hypothesis from data, the scientist may begin with a hypothesis and deduce conclusions—general statements or particular predictions—from it. This is *hypothetico-deduction* (H-D). Einstein reasoned hypothetico-deductively in constructing his special theory of relativity. He was committed to two fundamental principles: relativity (there is no absolute reference-frame; all motion is relative to an observer) and operational definition (scientific concepts should be defined in terms of observable phenomena). From the first principle he derived the paradoxical conclusion that the speed of light is constant, and found this conclusion confirmed by the Michelson-Morley experiment. From the second he deduced the paradoxical conclusion that measurements of simultaneity and distance are relative, and he found this conclusion incorporated in the Lorentz transformations (fundamental equations proposed by the Dutch physicist Hendrik Lorentz).

The scientist reasons *inductively* when he infers a general regularity from statements of particular instances. Early in the nineteenth century the French scientist Joseph Gay-Lussac reasoned inductively to the law that gases combine in simple ratios. He carried out experiments with various gases such as hydrogen and oxygen, fluoboric acid gas, and ammonia. From the fact that hydrogen and oxygen combine in a simple ratio and that various acid gases do so when combined with ammonia (and from certain theoretical assumptions, including the notion that gases, by virtue of their molecular structure, should obey simple laws), Gay-Lussac concluded that *all* gases combine in simple ratios.

Analogical reasoning is employed when the scientist arrives at a hypothesis by seeing an analogy between apparently unrelated phenomena. Darwin reasoned to part of his idea of natural selection from an analogy between population pressure among human beings (Malthus) and the survival of species in nature. Kekulé reasoned to his theory of the ring structure of the benzene molecule when he perceived an analogy between a snake holding its tail in its mouth and the arrangement of the carbon atoms in the benzene molecule.

Testing a Hypothesis Experimentally

When the scientist has formulated his hypothesis, he tests it by deducing its implications in the form of predictions and comparing them with the results of observations or experiments.

The reasoning behind the experimental test of a hypothesis is as follows. When the scientist seeks to establish a connection between two sets of events, he usually tries to show that one of these sets is the cause of the other. That is, he seeks to demonstrate that an event of one sort, *A*, is always accompanied by an event of another sort, *B*, and that an instance of *B* never occurs unless an instance of *A* also occurs. *A*, then, causes *B* when *A* must be present if *B* is to happen and when, with *A* absent, *B* never happens.

To show that one event is the cause of another is by no means easy. Any event, *B*, in nature usually occurs in combination with so many other events that it is hard to tell which of them is the cause of *B* and which accompany *B* by chance. One way of finding out is to create a situation in which we ourselves control the accompanying events (or conditions). We then can manipulate them one after another to ascertain which produces *B* and which do not. To do this, we produce a change in the condition we think is the cause of *B*, while keeping all other conditions from changing. If we then observe a change in the event, *B*, that follows, we may attribute it to the change that we ourselves have produced. This is our first experiment. We may then carry out a second experiment by varying some condition that we think has no significant influence on *B*, while holding unchanged the condition that we think produces *B*. If we then observe no significant change in *B*, we infer that *B* is affected only by a change in the original condition and not by a change in the other. When a scientist conducts an experimental test, he first deduces what his hypothesis implies for a given experimental situation and then manipulates the situation to see whether he is right. Consider one of the most famous of all experimental tests, the experiment in vaccination carried out at Pouilly le Fort, France, by Louis Pasteur in 1881. Pasteur wished to test the hypothesis that vaccinating an animal with attenuated (weakened) anthrax bacteria would make the animal immune to the disease of anthrax itself. Given 60 sheep by the local agricultural society, he divided the animals into three groups: (1) a control group of 10 sheep that were to receive no treatment whatever; (2) an experimental group of 25 that would be vaccinated and then inoculated with a highly virulent culture of the anthrax germ; and (3) another group of 25 that would not be vaccinated but would receive the same virulent dose of the germ. The experimental group would be vaccinated twice with anthrax bacteria of decreasing attenuation, at intervals of 10 to 15 days, and would be injected with a virulent dose of the germ 12 to 15 days later. Pasteur predicted that the 25 vaccinated sheep would all survive, and the 25 unvaccinated ones would die. The survivors would then be compared with the 10 control sheep to show that vaccination had done them no harm.

The first vaccinations were carried out on May 5 before a large crowd and were followed later by the second set of vaccinations and the administration of the germ itself. On June 2 Pasteur arrived to see the results. His predictions were fulfilled to the letter, as an eyewitness has described:

> When Pasteur arrived at two o'clock in the afternoon . . . accompanied by his young collaborators, a murmur of applause arose, which soon became loud acclamation, bursting from all lips. Delegates from the Agricultural Society of Melun, from medical societies, veterinary societies, from the Central Council of Hygiene of Seine et Marne, journalists, small farmers who had been divided in their minds by laudatory or injurious newspaper articles—all were there. The carcasses of twenty-two unvaccinated sheep were lying side by side; two others were breathing their last; the last survivors of the sacrificed lot showed all the characteristic symptoms [of anthrax]. All the vaccinated sheep were in perfect health. The one remaining unvaccinated sheep died that same night.

Experiments also may be carried out for fact-finding purposes with no hypothesis involved. One set of such experiments led, quite unexpectedly, to Rutherford's theory of the atomic nucleus. One day in 1909 Rutherford had a student, Ernest Marsden, try scattering alpha particles through a large angle, because Rutherford did not think it could be done. He described the result in one of his last lectures:

> Then I remember two or three days later Geiger [an associate] coming to me in great excitement and saying, "We have been able to get some of the alpha particles coming backwards. . . ." It was quite the most incredible event that has ever happened to me in my life. It was almost as incredible as if you fired a 15-inch shell at a piece of tissue paper and it came back and hit you.

But experimentation is not a *sine qua non* of scientific testing. Often it is physically impossible for the scientist to manipulate the circumstances of the phenomenon he wishes to explain or to do so without distorting them. On other occasions an experiment is not only physically impossible but logically inappropriate. The scientist may wish to explain some past event such as the event or events indicated by the presence of certain fossils in a stratum. Since this event is nonrecurring, it cannot be repeated in an experiment.

Some hypotheses cannot be tested decisively even by observation. Take Darwinian evolution. Although there is much evidence that species evolve, evolution itself is almost impossible to observe, for a variation only establishes itself over the course of many generations, and we cannot be around to watch the entire process. Nevertheless, the theory is taken to be well confirmed, not because it is decisively testable but because it unifies and renders intelligible a great many data that could not be understood without it.

A single successful prediction serves as a first confirmation of a hypothesis but does not make it reliable. That status normally is achieved only after the hypothesis has been tested and confirmed by a variety of scientists under a variety of conditions. A hypothesis, for example, that an agent such as nicotine or cyclamate is cancer-producing usually is tested in a number of laboratories against a range of animal species to determine whether the hypothesis applies to one species, or several, or all. If the hypothesis is confirmed by a variety of tests, it is regarded as reliable; and scientists may then begin to look for a mechanism to account for the correlation observed.

The Scientific Method in Perspective

The method I have described may seem hopelessly ideal. To the scientist, trying one hypothesis after another, the quest for a solution often feels like a series of setbacks leading nowhere. Yet the process of hypothesis, inference, test, and feedback is going on all the same. The scientist's moods may change, and so may his ideas, but not the essence of what he is doing. Looking back from the solution, the way through a problem may seem devious, but it is not devious in relation to the ignorance from which the scientist started. The method we have described is precisely that of feeling one's way in uncertainty.

I am not surprised that the scientist often is unaware of the method he uses, for as I have said, it is only an extension and refinement of the process of investigation followed in everyday life. As Max Planck explains, "Scientific reasoning does not differ from ordinary everyday thinking in kind, but merely in degree of refinement and accuracy, more or less as the performance of the microscope differs from that of the everyday eye." The scientist generally tries harder than the layman to screen out personal prejudice and check for possible error. He seeks to make his assumptions explicit and attends to the work of others in his field. He reports his findings more accurately and makes predictions that can in principle be tested precisely. In all these respects he improves on the layman but does not eclipse him. In fields where laymen have experience they may have more insight than highly trained scientists. Farmers or fishermen, for instance, often can predict local weather more accurately than meteorologists (in part, of course, because meteorology has not yet become as exact a science as physics).

Summary

By "scientific method" we mean the rational structure of those scientific investigations in which hypotheses are formed and tested. This structure is much like that of everyday thoughtful problem-solving. Hypothesis, inference, test, and feedback are the core of the structure. The scientist usually begins by noticing an anomalous fact or an inconsistency in theory and posing the discrepancy as a problem. After further exploration he proposes a hypothesis from which he deduces predictions. As a rule he tests the predictions and publishes the hypothesis if he finds them confirmed. If they are refuted, he usually alters the hypothesis, or invents another one, and tries again. This process is self-corrective. By eliminating incorrect hypotheses, the scientist narrows the search for the correct one.

This method is combined with general operations such as observation and measurement, and with various techniques differing from specialty to specialty. Scientific observation, often controlled by a hypothesis and aided by instruments, is more systematic and precise than its everyday counterpart. The data obtained by measurement and observation usually are theory-laden, and they are objective to the extent that they can be replicated by suitably qualified scientists. Scientific research, however, is focused as a rule on problems.

The crucial step in the research cycle is the invention of a hypothesis. The ideal hypothesis is precise and testable, accounts for the known facts, and predicts at least one new fact. The scientist usually tests his hypothesis by deducing its implications and then carrying out observations or experiments to see whether the implications correspond with the facts. Sometimes the scientist will run through a series of hypotheses and tests until he finds a hypothesis that he regards as satisfactory. Many hypotheses are tested experimentally. An experiment enables the scientist to manipulate the conditions accompanying a phenomenon until he discovers which of them cause it.

Not a few hypotheses are born in a moment of insight. Does this mean that the process of hypothesis-formation is nonrational? Not at all. Intuition seems to be only the condensation of a process of reasoning that normally would take longer and that in principle, if not always in practice, can be reconstructed later. Indeed, several modes of reasoning to a hypothesis have been identified: retroduction, hypothetico-deduction, induction, and reasoning by analogy.

The scientific method is not only intrinsically rational; it is a refinement of everyday reasoning. The scientist has received a more specialized training than the layman, but his thinking is not fundamentally different.

QUESTIONS FOR DISCUSSION

1. In the second paragraph the author defines "original research projects" as "investigations in which a hypothesis is formed." In what sense is research involving a hypothesis "original"?
2. What does the author mean when he says on p. 15 that "both facts and data are 'theory-laden'"? What are some of the implications for scientists of this notion?
3. What does it mean to say that scientists deduce predictions from their hypotheses/theories? Why must a good theory go beyond the known facts and predict some fact or facts hitherto unknown?
4. In his discussion of the research cycle, Kneller says it begins with either an empirical or conceptual problem. Explain the difference. Does this distinction help us understand the essential character of scientific enquiry?

HENRY H. BAUER

The So-called Scientific Method*

Born in Vienna, Henry H. Bauer was educated at the University of Sydney, where he received a Ph.D. in Chemistry in 1956. Bauer has taught at several universities and is currently Professor of Chemistry and Science Studies at Virginia Polytechnic Institute. In addition to chemistry, his interests include distinguishing science from pseudo-science and examining interactions of science and society. These latter interests have resulted in several books directed to a more general audience, including The Enigma of Loch Ness: Making Sense of a Mystery *(1986) and* Scientific Literacy and the Myth of the Scientific Method *(1992), from which the following selection is taken.*

In this piece, Bauer calls into question the traditional notion that science proceeds by means of a single, unambiguous "scientific method." For him, this traditional view fails to take into account the variety of approaches within any given scientific discipline, to say nothing of the important differences in method among the various disciplines. This essay can be usefully read along with those by Kneller and Renouf (this Part).

It is widely believed that the essence of science is its method. The definition used in surveys of scientific literacy expresses commonly held notions of what the scientific method is: systematic, controlled observation or experiment whose results lead to hypotheses, which are found valid or invalid through further work, leading to theories that are reliable because they were arrived at with initial open-mindedness and continual critical skepticism.

*Originally appeared in *Scientific Literacy and the Myth of the Scientific Method* by Henry H. Bauer.

One universally acknowledged source of this view is Francis Bacon (1561–1626); knowledge, he said, should come by generalizing from what one actually observes in the world—by contrast with the classic, Aristotelian approach of deducing with logical rigor from axiomatic first principles. But over centuries of argument and refinement, it has become clear that Baconian, inductive work could never establish truly certain knowledge: the *next* observation might force a different view or theory to be adopted even if the previous million observations had not—for example, just because the first million swans observed were white, one could not guarantee that the next one would be white also.

Distinctions were then suggested between observable and nonobservable things, it being supposed that at least one could be certain about observables, even if knowledge about nonobservables was inherently less reliable. Karl Popper introduced the influential insight that theories could never be positively proven to be true, whereas some theories could sometimes be definitively disproved; so, he suggested, to be scientific meant to deal in theories that could—at least in principle—be falsified.

Certain other significant issues in the philosophy of science will be taken up later, for example, the "theory-ladenness of facts"—to what degree do we observe what we believe we shall observe, by contrast with what may (or may not) be really there? For our present purpose, it is sufficient to recognize that these are the salient acknowledged elements of the popular view of being scientifically methodical: empirical, pragmatic, open-minded, skeptical, sensitive to possibilities of falsifying; thereby establishing objective facts leading to hypotheses, to laws, to theories; and incessantly reaching out for new knowledge, new discoveries, new facts, and new theories.

The burden of the following will be how misleading this view—which I shall call "the myth of the scientific method"—is in many specific directions; how incapable it is of explaining what happens in science; how it is worse than useless as a guide to what society ought to do about science and technology.

Are Chemists Not Scientists?

The scientific method is empirical. Scientific theories result when observation confirms tentative hypotheses. When the evidence speaks against them, hypotheses are falsified and therefore discarded.

One of my fellow graduate students in chemistry at the University of Sydney many years ago was trying to calculate certain properties of molecules, and he was the first to try to take account of one relatively subtle factor. Unfortunately, his calculations turned out to differ from the experimental values by more than earlier calculations had.

According to the Method for being Scientific, Dave should have considered his calculations falsified and tried another tack. Instead, he and his faculty advisor *ignored the comparison with experiment!* They were both mighty pleased with the progress Dave had made. He graduated top of our class, not much later he was on the faculty at Oxford, and soon after that he was a Fellow of the Royal Society.

Dave is far from alone among chemists in trusting theory more than experiment. A few years ago, a review article in *Science* listed many instances in which calculations had been right while experiment had been wrong: for the energy required to break molecules of hydrogen into atoms; for the geometry and energy content of CH_2 (the unstable "molecule" in which two hydrogen atoms are linked to a carbon atom); for the energy required to replace the hydrogen atom in HF (hydrogen fluoride) by a different hydrogen atom; and for others as well. The author, H. F. Schaefer, argued that good calculations—in other words, theory—may quite often be more reliable than experiments . . .

That is the view of one who is a theoretician, of course. You do not have to be long in a chemistry department to learn that chemists are no homogeneous tribe but rather a (sometimes uneasy) confederation of several distinct tribes: the analytikers, the inorganikers, the organikers, and the physical chemists are almost universally recognized to be distinctly different; and among or within these, or occasionally as separate tribes in their own right, there are electrochemists, polymer chemists, theoretical chemists, and others as well. And there are further subdivisions still: for instance, within many of these tribes, into experimentists and theorists.

Naturally, each tribe and subtribe thinks its own way of doing things to be the best way, the *scientific* way. So theorists tend to believe that experimental evidence is important only insofar as it suggests new theory; and if experiment and theory happen not to agree, the theorists will often prefer to believe the theory rather than the (experimental) evidence. Experimentalists, on the other hand, regard that as perverse; they know it is observation and experiment that teach us about how the world works, theories being only devices that make it easier to remember the facts.

Both sides have something of a point (though they rarely manage to get much beyond it). Taking for the moment the side of the theorists, it is unquestionably the case that failure to go beyond what experiment shows can mean that discoveries are missed. A dramatic instance is that of the structure of DNA (the molecules that convey hereditary information), whose elucidation is generally credited to James Watson and Francis Crick. One crucial bit of information was that DNA contains equal amounts of the substances A (adenine) and T (thymine), and also equal amounts of G (guanine) and C (cytosine). Those equalities had been indicated by the lengthy, painstaking experimental work of Erwin Chargaff, who has in more recent years made abundantly clear his belief that Watson, Crick, and the Nobel Prize Committee did not give him due credit. But when you look at Chargaff's publications, what you find are tables like those in Table 1, in which the amounts of A and T, and of G and C, are only *approximately* equal (or, what amounts to the same thing, in which the ratio A:G is only *about* the same as T:C). Chargaff wrote about his observations: "The results serve to disprove the tetranucleotide hypothesis. It is, however, noteworthy—whether this is more than accidental, cannot yet be said—that in all desoxypentose nucleic acids examined thus far the molar ratios of total purines to total pyrimidines, and also of adenine to thymine and of guanine to cytosine, were not far from 1."

By refusing to stick his neck out beyond the actual results and say plainly that they mean exact equality and hence some sort of pairing in the molecular structure, Chargaff may have missed out on a share in the Nobel Prize. Watson and Crick, on the other hand, were speculating and theorizing about the molecular nature and biological functions of DNA; and they postulated a structure in which the equalities are *exactly* 1–deviations found from that in actual practice could be regarded as experimental

	In Ox Thymus			In Ox Spleen		In Human Sperm		In Human	In Avian	In Yeast	
	1	2	3	1	2	1	2	Thymus	Tubercle Bacilli	1	2
A(adenine)	0.26	0.28	0.30	0.25	0.26	0.29	0.27	0.28	0.12	0.24	0.30
G(guanine)	0.21	0.24	0.22	0.20	0.21	0.18	0.17	0.19	0.28	0.14	0.18
C(cytosine)	0.16	0.18	0.17	0.15	0.17	0.18	0.18	0.16	0.26	0.13	0.15
T(thymine)	0.25	0.24	0.25	0.24	0.24	0.31	0.30	0.28	0.11	0.25	0.29

	In Ox Thymus	In Ox Spleen	In Ox (average)	In Human Thymus	In Human Sperm	In Human Liver		In Human (average)	In Yeast
						Normal	Cancer		
A/G	1.3	1.2	1.29 (±0.13)	1.5	1.6	1.5	1.5	1.56 (±0.008)	1.72
T/C	1.4	1.5	1.43 (±0.03)	1.8	1.7	1.8	1.8	1.75 (±0.03)	1.9

Table 1. Erwin Chargaff's data about the chemical composition of DNA, taken from his "Chemical Specificity of Nucleic Acids and Mechanism of Their Enzymatic Degradation," *Experientia* 6 (1950): 201–40.

errors. Watson and Crick turned out to be (largely) right; so, once again, ideas or theory had turned out to be a better guide than raw data, to what it all means. Inevitably so, for raw uninterpreted data do not mean anything: meaning rests on interpretation.

Evidently, then, some of the most successful chemists have not practiced the proper scientific method, which is supposed to put evidence first and theorizing second.

Science is also supposed to seek new discoveries; but, it turns out, chemists often do not welcome new discoveries. For example, if you read about chemical reactions that oscillate periodically, you find that William C. Bray's discovery of such a reaction in 1921 was simply not believed. Some thirty years later, in 1951, a paper by B. P. Belousov on the same subject was rejected, the editor saying that the reported results were simply impossible. Finally in the 1970s these results came to be accepted, *but only after a theoretical treatment had shown how oscillations could come about.* Again, more heed had been paid to theory—which is to say to preconceived belief—than to plain empirical fact.

Is Anyone a Scientist?

I should make quite plain that I do not *really* want to say that chemists are not proper scientists. What I just did with chemists (because I happen to know them best, being one myself) could equally be done with astronomers, or biologists, or geologists, or physicists, or with any of the other tribes within science. The point I do wish to make is that purportedly authoritative pronouncements as well as popular ideas about how science works are very seriously mistaken. One can find innumerable examples in all the sciences where theory was believed in the face of apparent evidence to the contrary; one can even find such an approach explicitly defended by eminent scientists—for example, the physicist Sir Arthur Eddington: "it is also a good rule not to put overmuch confidence in the observational results that are put forward until they have been confirmed by theory."

Even worse, theory—which, remember, is preconceived belief—may cause scientists to think they are observing things that in actuality do not exist, like the canals of Mars. And all the sciences offer instances where major new discoveries have not been accepted for quite a while because they ran counter to existing beliefs—consider the discoveries of Hermann Helmholtz and Max Planck, of Joseph Lister in medicine, of Oliver Heaviside in mathematical physics, Thomas Young's wave theory of light, and the cases of Louis Pasteur and Gregor Mendel and Svante Arrhenius and on and on and on.

So the classical and common view of science misconceives the actual relationship between theories and facts; and (consequently, inevitably) it misconceives the nature of the scientific method—the things that scientists actually do. It misconceives the behavior of science and of scientists in the face of surprising discoveries; and it misconceives much else about science, about technology, and about their interaction with one another and with the wider society.

An important misconception is implicit in the very use of the terms "science," "scientists," "scientific." To talk of scientists is to imply that astronomers, biologists, chemists, geologists, and physicists are all somehow much the same in some significant respect. To talk of science is to imply that astronomy, biology, chemistry, geology, and physics are all much the same sort of things. When there is talk about being scientific, it is commonly implied that one can be that, scientifically methodical, irrespective of the particular nature of what is being done; that one can be scientific about anything, like canvassing for new members for a bridge club: "Westchester County . . . has a tremendous program . . . working on memberships scientifically—how to get the people and how to keep them."

As soon as one looks in any depth, however, it becomes less and less clear what is really the same about astronomy and biology, say, or about what astronomers do and what biologists do. Sure enough, both astronomy and biology (and the other sciences as well) have to do with the study of selected aspects of nature. Sure enough, their findings are always subject to the commands of reality: false results are discarded (sooner or later, as their falsity becomes sufficiently obvious). Sure enough, each of the sciences now offers impressively detailed, coherent, and reliable insights, far more than they did fifty years ago, vastly more than a century ago, almost unrecognizably more than two centuries ago.

But there, or about there, the identity among the sciences comes to an end. The diversity among them includes that they vary in the degree to which they use mathematics: physics and astronomy cannot do without high mathematics, whereas much of biology or geology needs little more than arithmetic, and various bits of chemistry fall into one or the other of those categories. Though this diversity is commonly acknowledged, not generally recognized is the degree to which that difference entails other significant differences of practice: whether or not quantification is seen as the ultimate aim, for example, and whether or not mathematics is a required part of a student's initial training, and whether or not one comes to equate "quantitative" with "scientific" or "good."

Again, the distinction between observational and experimental science is a commonplace, but appreciation of the corollaries is not. Yet the way in which observational astronomers work has little in common with the way experimental chemists work: they differ in the sorts of funding they apply for (telescope time or chunks of actual money), in their reliance on graduate students (optional as opposed to essential), in the frequency with which they are expected to publish, in the way their peers interpret the significance of articles published with many coauthors, and in all sorts of other ways as well. Observation is so much more at the mercy of nature than is experiment that meaningful distinctions are obscured when—as is so often done—the two are lumped artlessly together as alternative but somehow equivalent modes of being empirical.

Many other consequential distinctions among the sciences are less frequently remarked. For example, whereas astronomy and biology and geology are fundamentally and inherently concerned with large-scale change that seems always to have gone in the same direction, chemistry and physics are not. Astronomy has to deal with the evolution of the universe, the birth and development and death of stars; biology and geology seek to account for the evolution of living things and of the Earth. But physics and chemistry share no such concern with inherent, directional change: they delight, by contrast, in the discovery of permanent relationships, and they do experiments in which time is just another controllable factor. Again, astronomy and biology and geology are, by and large, observational sciences, studying whatever nature presents them with, whereas chemistry and physics, by and large elemental sciences, can decide what to study, within increasingly wide limits—to the extent of making materials and arranging conditions that nature never before knew.

With these and other differences among sciences come far-reaching differences in attitude and method on the part of those who do the science, differences not often explicitly recognized. For example, chemists and physicists do not mean quite the same thing when they call a thing "stable": physicists mean that it is stable for all time, that it is in its lowest state of energy and will remain there until disturbed by another object or a force, whereas chemists mean that the thing does not by or of itself change into something else *at a noticeable rate* in a normal environment. The differences among adepts of the various sciences go beyond matters of theory, method, and vocabulary to subtler habits of thought and even to customs of behavior, to such an extent that the differences among the sciences, not only between the sciences and the humanities, can aptly be described as cultural ones: they involve a great deal more than just knowing about separate and distinct aspects of nature. Thus biologists and experimental physicists use visual imagery more than do theoretical physicists. Much theoretical speculation and argumentation over very few facts is commonplace in paleoanthropology or in astronomy but not in chemistry or in geology. Physicists look to crucial experiments to decide among theories at one fell swoop, whereas astronomers are used to waiting for long periods of time for the accumulation of data to bring an end to the speculation. Nobel Prizes in physics have been awarded about twice as often for experimental novelties as for theoretical ones, but in chemistry, experimentalists have been so honored five or six times as often as have theorists. Eminent physicists were found to feel pressed for time more than were eminent biologists; and the physicists gave up research in favor of administration at an earlier age. In matters of politics, physicists are considerably more liberal, on the average, than other scientists. Rates of divorce were found to be three times as great among biologists as among physicists (some decades ago, one should perhaps stress, when all the rates were lower than they now are). Though there seems not to have been any systematic study made of the matter, illustrations such as these are readily enough found to make the point that sciences differ among one another along many dimensions and not merely in commanding "knowledge" about separate pieces of the natural world.

Once the point is recognized, reasons can readily be suggested for some of these variations. As science developed over the last few centuries, the growth of knowledge demanded specialization. But the specialization unavoidably and soon became much more than a concern with distinct sets of phenomena. Those who studied some things found that they progressed best by taking more note of theory, whereas others found themselves going astray if they ventured too far from observation—and so some specialties came to understand that experimental evidence should not be accepted until it has been confirmed by theory, whereas most sciences and most scientists at least claim to believe the opposite. Each science—and to a degree each specialization within each science—has thus come to be an idiosyncratic blend of theorizing and empiricism; and that brings inevitably with it distinct notions about what knowledge (in general!) is and about the degree to which knowledge can be said to be "certain." In turn, disparate views about the nature of knowledge lead to different judgments about what might be interesting, valuable, fruitful to study. What we call "science" nowadays encompasses a wide range not only of knowledge but also of diverse views about the nature of knowledge.

Consider, as a last illustration, geology and physics. Physics is very oriented toward theory: one learns physics as a set of mathematically formulated laws more than as a set of observed phenomena; theory serves as a substitute for individual facts. Most physicists asked about the attraction between two of the planets will look up their masses and do a calculation; it would not occur to them to seek direct observational data. Geology, by contrast, is taught primarily through description—of minerals, geographic and geophysical features, strata, and fossils; theory in geology is less specific than in physics and serves to explain after the fact and not as a substitute for individual facts. Naturally, then, physicists tend to regard quantitative theory as the epitome of science and of scientificity; and, secretly or not so secretly, they see geology and geologists as somewhat less than highly scientific. So, too, physicists have learned that it is possible to find distinct, single causes for the variety of phenomena with which they deal, the phenomena themselves being identifiably and distinctly discrete. And for these reasons, and also because they can control all the relevant factors, physicists know that they can perform "crucial experiments" that compel nature to deliver definite answers. Geologists, on the other hand, learn that their phenomena overlap one another, that diverse "causes" conjointly produce any given geological circumstances and that the most scientific approach is not that of seeking crucial tests but that of "multiple working hypotheses," for in geology one must, over long periods of time, be willing to countenance the possibility that any one of several competing explanations may ultimately turn out to be the best.

In view of such differences, it should not be surprising, for example, that it was a physicist who pushed most dogmatically the view that the dinosaurs were killed off in a discrete event by the simple cause of the collision with Earth of an asteroid, the impact of which remains demonstrable through the layer of iridium deposited at that time at the boundary between the Cretaceous and the Tertiary strata. To paleontologists, however, that seems absurdly oversimplified. For them, the layer of iridium-rich material is rather or partly an absence of carbonate sedimentation from the oceans during long eras in which almost no limestone was formed; the extinction of the dinosaurs itself is not seen as an event but as part of the change, over the course of millions of years, in the number of species as well as of individual dinosaurs; and, moreover, that extinction is looked at as only one of a number of occasions within geologic time when the diversity of living species decreased markedly, making it a statistical fluctuation within the perpetual flux of the appearance and disappearance of species, not a discrete, unusual individual occurrence.

Thus geologists and physicists tend to approach even scientific problems in disparate ways. They learn differently what it is to be scientific, what the scientific method is; and so too do chemists and biologists and other scientists come to different and even contradictory views of what science is. Yet these characteristic differences are but little recognized, and the misconception remains widespread that there exists a single method whose utilization marks the whole of science. In point of fact, as just illustrated, there is not any single thing that one can usefully and globally call science; rather, there are many different sorts of science. Once one has said that science is the study of nature, and that scientific knowledge is valid only so long as it is not contradicted by nature, one has said essentially all that is truly common, without qualification, among all the sciences. Beyond that, one finds nothing but variation: in the degree of weight put upon evidence in comparison with theory, in the ease with which data can be gathered, and in innumerable other details.

Diverse Aspects of Science

Some of the ways in which diverse bits of science differ from one another are indicated in Table 2. In what follows, consequences of this diversity will be illustrated in an oversimplified way by comparing or contrasting whole sciences with one another; but these distinctions actually characterize most faithfully only quite small bits of science within any of the major disciplines. Thus chemistry as a whole is relatively mature, data rich, experimental, and data driven, but many bits of chemistry are not mature (the chemistry of metal clusters, for instance, or of catalysts); quantum chemistry is a recognized subdiscipline that is notably theory driven, not data driven; and so on.

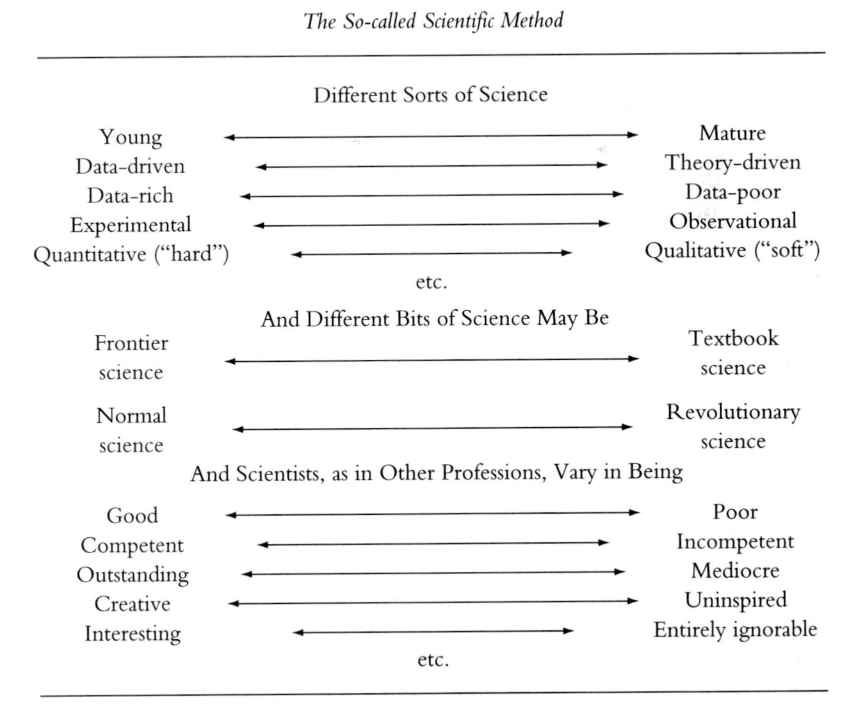

The So-called Scientific Method

	Different Sorts of Science	
Young	←→	Mature
Data-driven	←→	Theory-driven
Data-rich	←→	Data-poor
Experimental	←→	Observational
Quantitative ("hard")	←→	Qualitative ("soft")
	etc.	
	And Different Bits of Science May Be	
Frontier science	←→	Textbook science
Normal science	←→	Revolutionary science
	And Scientists, as in Other Professions, Vary in Being	
Good	←→	Poor
Competent	←→	Incompetent
Outstanding	←→	Mediocre
Creative	←→	Uninspired
Interesting	←→	Entirely ignorable
	etc.	

Table 2. Scientific activity displays a very wide range of characteristics.

Some of the variations among bits of science have to do with their individual degrees of maturity. Physics is as a whole the most mature of the major sciences, and it seems plausible that this is a cause underlying the fact that physics also has the greatest degree of unanimity within its ranks: hardly anyone within the discipline questions relativity or quantum mechanics, or that the salient task for the discipline is to achieve the theoretical Grand Unification of the Four Forces. There is little ferment over what should be taught in physics courses, or how. The professional journals stand in well-recognized differentiation of function and hierarchy of status. The professional societies work quietly and without fuss. By contrast, the young computer science is in ferment over most of those things: what the most pressing goals of the field are; whether the subject itself is more akin to engineering or to science; how its practitioners should be trained; what constitutes professional publication; and so forth. Physicists look to governments for their major funds, whereas computer scientists look to their own institutions and to the computer industry. Physicists display the patina of an established aristocracy, whereas computer scientists exhibit characteristics of the nouveau riche.

It should be noted, however, that the theories commanded by a mature science are not necessarily more final and true than those commanded by a young science, though the assured behavior of the practitioners of the mature might lead one to think so. Physics for centuries has been the most mature among the sciences; and physicists periodically lapse into the belief that all the important principles of their subject have been discovered and that what remains is only to fill in the details. That was the case, for instance, circa 1870; but it was followed by disillusionment and then some of the most revolutionary bits of science: radioactivity, revealing that atoms are not stable and indestructible; relativity, altering drastically the notions of time and space and gravitation and motion; quantum phenomena, and the paradoxical view that some things could behave at times like particles but at other times like waves. Some of the most fundamental theoretical principles were replaced. (Yet, it should be noted, the vast majority of *knowledge*, in contrast to *theories*, in physics remained intact. A piece of uranium ore will always cause a photographic plate to become exposed, even through its protective black paper covering, and that tells us something definite about the external world; though the terms in which we try to explain that, connecting it with other phenomena, are always subject to some degree of change. That burning means combining with oxygen and not the release of phlogiston was the central realization of the most significant resolution in chemistry; but an enormous amount of what chemists know about combustion—which substances will burn in the presence of each other—is the same now as it was before that revolution. Thus much of the knowledge of what happens in nature, together with a great deal of the explanation that ties those phenomena together, remains essentially unaffected by apparently revolutionary shifts in theory.)

Again, mature science is not necessarily data rich; desired data are not necessarily acquired easily. To a possibly (but by no means certainly) apocryphal chemist is attributed the statement that he could not be bothered wading through the literature because he could get any answer he wanted more quickly by doing the experiment. Though physics is a more mature science than is chemistry, physicists would not venture that jocularity—though they might well say that they do not look it up because they can calculate it. Experiments in much of chemistry are more readily performed than in most of physics.

Much of chemistry is indeed data rich: feasible experiments can deliver a wealth of results in little time, and the chemical literature brims with data about millions of substances, about their preparation and properties. Consequently, chemistry has developed its idiosyncratic judgment about what and how much a chemist ought to accomplish. On the one hand, chemists publish—and are expected by their peers to publish—more articles than, say, geologists; on the other hand, chemists derogate those who do nothing but "turn the crank"—who use a given technique or instrument to derive data from a succession of different substances or reactions—because that is so easy a thing to do. Again because data are generated in such massive amounts, chemistry was the leader in developing techniques for trying to control the exploding literature, through "abstracts" journals and then computerized on-line abstracts. And again because so much can be done, chemists desperately want to have many co-workers to push their research forward, and much about university departments of chemistry can be understood as flowing from the overriding need to recruit plenty of graduate students, "pairs of hands" to carry out the necessary experiments. Thus there is related the story of X, a certain graduate student in chemistry who had the temerity to leave town with the wife of his supervising professor, Y. And X had the further temerity, a year or so later, to ask Y whether he could return to complete the work for his degree. As Y later said to a colleague, "Of course I agreed. He's first rate, and as you know, good graduate students are hard to come by while wives are a dime a dozen." One would not be quite as ready to believe that story if told about an astronomer or a geologist, for many excellent careers in astronomy and in geology, but few in chemistry, have been made without reliance on the labor of co-workers.

In data-rich fields, theory has a certain down-to-earth quality: speculation is fettered by the ease with which it can be disproved. But in data-poor fields, extremely tenuous chains of speculation are indulged. Thus cosmologists are notorious for theorizing whose equivalent on other subjects would be dismissed as science fiction: they are free, for example, to publish about what things may have been like before the big bang that started everything we know about; astrophysicists and cosmologists accept as conceivable the interpretation that certain observations are of objects producing inconceivable amounts of energy by means of inconceivable processes; those searching for extraterrestrial intelligence have published copiously, in almost total absence of data and in complete absence of any direct data; paleoanthropologists construct whole new charts of human evolution whenever a new fossil is found. By contrast, geologists denied continental drift for decades, supported though it was by fitting coastlines and biogeographical distributions. Geologists are always faced with a complex richness of data that offers continuing challenges even to meaningful categorization, let alone explanation, and so geologists are used to waiting and waiting for explanatory schemes to be developed; there is no hurry for that, explanation is not (yet) so salient a part of geology, and they have plenty of useful and time-consuming things to do without indulging in grandiose theorizing.

Some scientists thus do a lot of speculating, whereas others do virtually none, and there is no warrant to call the one approach scientific and the other not. It is just the case that different aspects of nature yield to investigation at different rates and in different ways, and so scientists come to differ in all manner of things. Whenever a generalization is made about science or about scientists, disregarding thereby the fact that there are so many distinct sorts of science, misconception is promulgated. What is true or fruitful for a field that is mature, data rich, and relatively quantitative (thermodynamics, say) is scientific for that specialty even though it may be entirely inappropriate and therefore unscientific for a field that is young, descriptive, and data poor (some bits of planetary science, say).

Again, though we use the single word "science" for both, *textbook* science is a very far cry from *frontier* science. What is in the texts is reliable. It is relatively uncolored by the personalities of those who originally conceived it. It is generally agreed to by almost all the experts. It is unlikely to need to be altered in the future, and in that unlikely event the alteration will likely be of limited extent. By contrast, science at the frontier is very unreliable: today's discovery often turns out tomorrow to have been an error. Frontier science often bears the stamp of its discoverer's persona; and it is often disputed by other experts. Frontier science and textbook science are about as different from one another as any two things can be, within the bounds that both are guesses about the nature of the real world.

Scientists Are Human

Finally, the common view of science as a unitary, monolithic enterprise fails to recognize how varied are the people who do it. Scientists are supposedly trained to judiciousness, objectivity, patience, and careful experimentation and observation; scant attention has been paid to how the practice of science is influenced by the fact that scientists, like all other human beings, vary in ability, competence, dedication, and honesty.

Indeed, thinking of science as using the scientific method portrays science as an activity that is highly unnatural: human beings are not by nature objective, judicious, disinterested, skeptical; rather, human beings jump to conclusions on flimsy evidence and then defend their beliefs irrationally. The widely held myth of the scientific method is one reason that scientists are often stereotyped as cold, even inhuman. Consider, for instance, what a celebrated humanist educator had to say:

> . . . teaching is an art, not a science. It seems to me very dangerous to apply the . . . methods of science to human beings. . . . a "scientific" relationship between human beings is bound to be inadequate and perhaps distorted. . . . to be orderly in planning. . . and precise in . . . dealing with facts . . . does not make . . . teaching "scientific.". . . A "scientifically" brought up child would be a pitiable monster. A "scientific" marriage would be only a thin and crippled version of a true marriage. A "scientific" friendship would be as cold as a chess problem. . . . Teaching is not like inducing a chemical reaction: it is much more like painting a picture or making a piece of music, or on a lower level like planting a garden or writing a friendly letter. You must throw your heart into it, you must realize that it cannot all be done by formulas.

Thus science, arguably the finest exemplar of human intellectual achievement, is made to appear at best a necessary evil. When science is pictured as so impersonal and ascetic an activity, how to understand that scientists *do* throw their hearts into their work, which also cannot and is not all done by formulas? The myth of the scientific method hinders recognition of the wonderful diversity of the sciences. It makes it impossible to understand the history of science and contemporary scientific activity, and it fosters the stereotype of the cold, inhuman, sometimes evil scientist.

QUESTIONS FOR DISCUSSION

1. In the opening paragraphs, Bauer describes the conventional view of the scientific method (Baconian/inductive: observation first, theory second). To what degree does Kneller (previous essay) subscribe to this conventional view?
2. On p. 29, Bauer says that "chemists often do not welcome new discoveries." He goes on to suggest that the same attitude towards discovery characterizes some other sciences. What factors might contribute to such an attitude?
3. In discussing the work of Watson and Crick, Bauer says that "meaning rests on interpretation." What does he mean by this? What are some of the implications of this notion for the scientist?
4. On p. 35 the author says, "Some scientists thus do a lot of speculating [theorizing], whereas others do virtually none. . ." What is the principal reason for this difference? What are some of its implications?
5. In the final paragraph, the author claims that science is "arguably the finest exemplar of human intellectual achievement." How might such a claim be justified?

HARRY COLLINS AND TREVOR PINCH

The Germs of Dissent: Louis Pasteur and the Origins of Life*

Harry Collins, Director of the Science Studies Centre at the University of Bath, and Trevor Pinch, Professor of Science and Technology Studies at Cornell University, have published widely on the social aspects of science and on the relationship between technology and human values.

In the following selection, taken from The Golem: What Everyone Should Know About Science, *Collins and Pinch challenge the notion that scientific understanding is necessarily founded upon the unambiguous results of neat and tidy experiments. Specifically, they argue that the pioneering work of Louis Pasteur advanced not on an unquestioning acceptance of experimental findings but frequently on a rejection of those findings based on intuitive understanding. In other words, the rigid application of "scientific method" as generally understood was not the hallmark of Pasteur's way of doing science. The reader might wish to place this particularized discussion in the context of Henry Bauer's more generalized treatment of scientific method in the preceding essay.*

Spontaneous Generation

"Spontaneous generation"is the name given to the doctrine that, under the right circumstances, life can form from dead matter. In a sense, nearly all of us believe in spontaneous generation, because we believe that life grew out of the primeval chemical slime covering the newly formed earth. This, however, is taken to be something that happened slowly, by chance, and once only in the history of the earth; it ought never to be seen in our lifetimes.

The question of the origin of life is, of course, as old as thought but, in the latter half of the nineteenth century, the debate raged within the scientific community. Could new life arise from sterile matter over and over again, in a few minutes or hours? When a flask of nutrients goes mouldy, is it because it has become contaminated with existing life forms which spread and multiply, or is it that life springs anew each time within the rich source of sustenance? It was a controversial issue, especially in nineteenth-century France because it touched upon deeply rooted religious and political sensibilities.

Our modern understanding of biochemistry, biology and the theory of evolution is founded on the idea that, aside from the peculiar conditions of pre-history, life can only arise from life. Like so many of our widespread scientific beliefs we tend to think that the modern view was formed rapidly and decisively; with a few brilliant experiments conducted in the 1860s, Louis Pasteur speedily defeated outright those who believed in spontaneous generation. But the route, though it might have been decisive in the end, was neither speedy nor straightforward. The opposition were crushed by political manoeuvering, by ridicule, and by Pasteur drawing farmers, brewers, and doctors to his cause. As late as 1910, an Englishman, Henry Bastian, believed in the spontaneous generation heresy. He died believing the evidence supported his view.

As in so many other scientific controversies, it was neither facts nor reason, but death and weight of numbers that defeated the minority view; facts and reasons, as always, were ambiguous. Nor should it be thought that it is just a matter of "those who will not see." Pasteur's most decisive victory—his defeat of fellow Frenchman Felix Pouchet, a respected naturalist from Rouen, in front of a commission set up by the French Academie des Sciences—rested on the biasses of the members and a great stroke of luck. Only in retrospect can we see how lucky Pasteur was.

The Nature of the Experiments

The best-known experiments to test spontaneous generation are simple in concept. Flasks of organic substances—milk, yeast water, infusions of hay, or whatever—are first boiled to destroy existing life. The steam drives out the air in the flasks. The flasks are then sealed. If the flasks remained sealed, no new life grows in them—this was uncontested. When air is readmitted, mould grows. Is it that the air contains a vital substance that permits the generation of new life, or is it that the air contains the already living germs—not metaphorical, but literal—of the mould. Pasteur claimed that mould would not grow if the newly admitted air was itself devoid of living organisms. He tried to show that the admission of sterile air to the flasks had no effect; only contaminated air gave rise to putrescence. His opponents claimed that the admission of even pure air was sufficient to allow the putrefaction of the organic fluids.

The elements of the experiment are, then:

1. one must know that the growth medium is sterile but has nutritive value;
2. one must know what happens when the flasks are opened; is sterile air being admitted or is contamination entering too?

Practical Answers to the Experimental Questions

Nowadays we believe we could answer those questions fairly easily, but in the nineteenth century the techniques for determining what was sterile and what was living were being established. Even what counted as life was not yet clear. It was widely accepted that life could not exist for long in a boiling fluid, so that boiling was an adequate means of sterilisation. Clearly, however, the medium could not be boiled dry without destroying its nutritive value. Even where the boiling was more gentle it might be that the "vegetative force" of the nutrient might have been destroyed along with the living organisms. What counted as sterile air was also unclear. The distribution of micro-organisms in the world around us, and their effect on the air which flowed into the flasks, was unknown.

Pasteur made attempts to observe germs directly. He looked through the microscope at dust filtered from the air and saw egg-like shapes that he took to be germs. But were they living, or were they merely dust? The exact nature of dust could only be established as part of the same process that established the nature of putrescence.

If germs in the air could not be directly observed, what could be used to indicate whether air admitted to a flask was contaminated or not? Air could be passed through caustic potash or through sulphuric acid, it could be heated to a very high temperature or filtered through cotton wool in the attempt to remove from it all traces of life. Experiments in the early and middle part of the nineteenth century, using air passed through acids or alkalis, heated or filtered, were suggestive, but never decisive. Though in most cases admission of air treated in this way did not cause sterilised fluids to corrupt, putrescence occurred in enough cases to allow the spontaneous generation hypothesis to live on. In any case, where the treatment of the air was extreme, it might have been that the vital component which engendered life had been destroyed, rendering the experiment as void as the air.

Air could have been taken from different places—high in the mountains, or low, near to cultivated fields—in the expectation that the extent of microbial contamination would differ. To establish the connection between dust and germs, other methods of filtration could be used. Pasteur used "swan neck flasks" (see Figure 1). In these the neck was narrowed and bent so that dust entering would be caught on the damp walls of the orifice. Experiments were conducted in the cellars of the Paris Observatoire, because there the air lay sufficiently undisturbed for life-bearing dust to have settled. Later on, the British scientist, William Tyndall, stored air in grease-coated vessels to trap all the dust before admitting it to the presence of putrescible substances. For each apparently definitive result, however, another experimenter would find mould in what should have been a sterile flask. The kinds of arguments that the protagonists would make can be set out on a simple diagram.

Box 1 is the position of those who think they have done experiments that show that life *does* grow in pure air and believe in spontaneous generation. They think these experiments prove their thesis. Box 2 is the position of those who look at the same experiments but do not believe in spontaneous generation; they think there must have been something wrong with the experiment, for example, that the air was not really pure.

Box 4 represents the position of those who think they have done experiments showing that life *does not* grow in pure air and do not believe in spontaneous generation. They think the experiments prove their hypothesis. Box 3 is the position of those who look at the same experiments but do believe in spontaneous generation. They think there must have been something wrong with the air, for example, that its vital properties were destroyed in the purifying process.

		Believe in spontaneous generation	
		Yes	No
Life grows in apparently pure air	Yes	(1) Proves thesis	(2) Air accidentally contaminated
	No	(3) Air spoiled by treatment	(4) Proves thesis

Possible interpretations of spontaneous generation experiments.

There was a period in the 1860s when arguments of the type found in box 3 were important but this phase of the debate was relatively short-lived; it ended as the experimenters ceased to sterilise their air by artificial means and instead sought pure sources of air, or room temperature methods of "filtration." Arguments such as those found in box 2 were important for a longer time. They allowed Pasteur virtually to define all air that gave rise to life in the flasks as contaminated, whether he could show it directly or not. This is especially obvious in that part of his debate with Felix Pouchet concerning experiments using mercury, as we shall see.

The Pasteur–Pouchet Debate

One episode of the long debate between Pasteur and those who believed in spontaneous generation illustrates clearly many of the themes of this story. In this drama, the elderly (60-year-old) Felix Pouchet appears to serve the role of "foil" for the young (37-

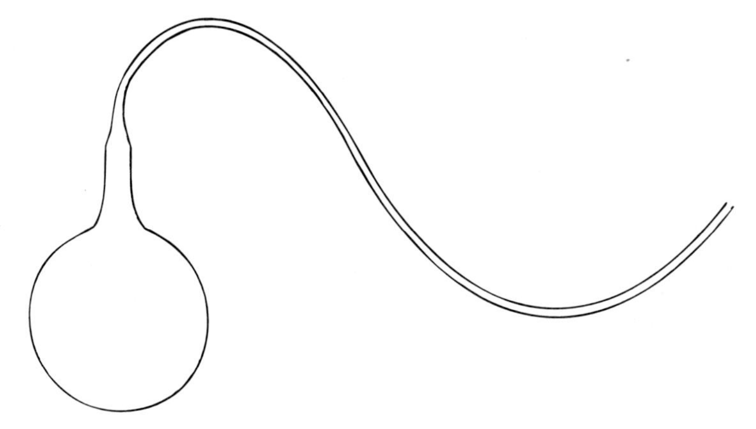

Figure 1. One of Pasteur's swan-neck flasks.

year-old) Pasteur's brilliant role as experimental scientist. Pasteur, there is no doubt, defeated Pouchet in a series of celebrated trials, but the retrospective and triumphalist account glosses over the ambiguities of the trials as they took place in real time.

As with all such experimental controversies, it is the details that are crucial. The argument between Pasteur and Pouchet concerned what happens when an infusion of hay—"hay tea," as one might say—which is sterilised by boiling, is exposed to the air. It is undisputed that the infusion goes mouldy—microscopic life forms grow upon its surface—but the usual question remained. Was this because air has life-generating properties or because air contains living "seeds" of mould?

Experiments "Under Mercury"

Pouchet was a believer in spontaneous generation. In his early experiments he prepared sterilised infusions of hay "under mercury"—to use the jargon. The method was to do the work with all vessels immersed in a mercury trough so that ordinary air could not enter. Specially prepared air could be introduced into the flask by bubbling through the mercury trough. This was the standard way of admitting various experimental gases into experimental spaces without admitting the ordinary air. In Pouchet's case it was purified air that was bubbled through the mercury. It was considered that purified air could be made by heating ordinary air, or by generating oxygen through the decomposition of an oxide; coincidentally this was often mercury oxide which gives off oxygen when heated. Invariably Pouchet found that when purified hay infusions were prepared under mercury, and exposed to pure air, organic life grew. It appeared then that, since all sources of existing life had been eliminated, the new life must have arisen spontaneously.

Pouchet started the debate with Pasteur by writing to him with the results of these experiments. Pasteur wrote back to Pouchet that he could not have been cautious enough in his experiments. "... in your recent experiments you have unwittingly introduced common [contaminated] air, so that the conclusions to which you have come are not founded on facts of irreproachable exactitude." Here, then, we see Pasteur using an argument of the type that is found in box 2 above. If Pouchet found life when he introduced sterilised air to sterilised hay infusions, then the air *must* have been contaminated.

Later, Pasteur was to claim that, although the hay infusion was sterile in these experiments, and the artificial air was equally devoid of life, it was the mercury that was contaminated with microorganisms—they were in the dust on the surface of the mercury—and this was the source of the germ.

This is interesting because it seems that the contaminated mercury hypothesis was necessary to explain some of Pasteur's own early results. He reported that in his own attempts to prevent the appearance of life by preparing infusions under mercury, he succeeded in only 10% of his experiments. Though, at the time, he did not know the source of the contamination, he did not accept these results as evidence in support of the spontaneous generation hypothesis. In his own words, he "... did not publish these experiments, for the consequences it was necessary to draw from them were too grave for me not to suspect some hidden cause of error in spite of the care I had taken to make them irreproachable." In other words, Pasteur was so committed in his opposition to spontaneous generation that he preferred to believe there was some unknown flaw in his work than to publish the results. He *defined* experiments that seemed to confirm spontaneous generation as unsuccessful, and vice versa. Later the notion of contaminated mercury replaced the "unknown flaw."

Looking back on the incident we must applaud Pasteur's foresight. He was right, of course, and had the courage of his convictions in sufficient degree to refuse to be swayed by what, on the face of it, was a contrary experimental indication. But it *was* foresight. It was not the neutral application of scientific method. If Pasteur, like Pouchet, had been supporting the wrong hypothesis we would now be calling his actions "dogged obstinacy in the face of the scientific facts." Perfect hindsight is a dangerous ally in the history of science. We shall not understand the Pasteur–Poucher debate as it was lived out unless we cut off our backward seeing faculty.

Flasks Exposed at Altitude

The business of the experiments under mercury was just the preliminary skirmish. The main debate began with Pasteur's experiments on flasks opened to the air at altitude, and Pouchet's rebuttal.

Pasteur prepared flasks with necks drawn out in a flame. He boiled an infusion of yeast and sealed the neck once the air had been driven out. If unopened, the contents would remain unchanged. He could then take the flasks and break the neck at various locations, allowing air to re-enter. To admit air in what ought to be germ-free locations, Pasteur would break the neck with a long pair of pincers which had been heated in a flame, while the flask was held above his head so as to avoid contamination from his clothes. Once the air from the chosen location had entered, Pasteur could once more seal the flask with a flame. Thus he prepared a series of flasks containing yeast infusions together with samples of air taken from different locations. He found that most flasks exposed in ordinary locations became mouldy, whereas those exposed high in the mountains rarely changed. Thus, of 20 flasks exposed at 2000 meters on a glacier in the French Alps, only one was affected.

In 1863, Pouchet challenged this finding. With two collaborators he travelled to the Pyrenees to repeat Pasteur's experiments. In their case, all eight of the flasks exposed at altitude were affected, suggesting that even uncontaminated air was sufficient to begin the life-forming process. Pouchet claimed that he had followed all of Pasteur's precautions, except that he had used a heated file instead of pincers to open the flasks.

Sins of Commission

In the highly centralised structure of French science in the mid-nineteenth century, scientific disputes were settled by appointing commissions of the Paris-based Academie des Sciences to decide on the matter. The outcomes of such commissions became the quasi-official view of the French scientific community. Two successive commissions looked into the spontaneous generation controversy. The first, set up before Pouchet's Pyrenean experiments, offered a prize to "him who by well-conducted experiments throws new light on the question of so-called spontaneous generation." By accident or design, all members of the commission were unsympathetic to Pouchet's ideas and some announced their conclusions before even examining the entries. Two of its members had already responded negatively to Pouchet's initial experiments and the others were well-known opponents of spontaneous generation. Pouchet withdrew from the competition, leaving Pasteur to receive the prize uncontested for a manuscript he had written in 1861, reporting his famous series of experiments demonstrating that decomposition of a variety of substances arose from air-borne germs.

The second commission was set up in 1864 in response to Pouchet's experiments in the Pyrenees. These experiments had aroused indignation in the Academie, most of whose members had considered the matter to be already settled. The new commission started out by making the challenging statement: "It is always possible in certain places to take a considerable quantity of air that has not been subjected to any physical or chemical change, and yet such air is insufficient to produce any alteration whatsoever in the most putrescible fluid." Poucher and his colleagues took up the challenge adding: "If a single one of our flasks remains unaltered, we shall loyally acknowledge our defeat."

The second commission too was composed of members whose views were known to be strongly and uniformly opposed to those of Pouchet. When he discovered its composition, Pouchet and his collaborators attempted to alter the terms of the test. They wanted to expand the scope of the experimental programme while Pasteur insisted that the test should depend narrowly upon whether the smallest quantity of air would always induce putrescence. All Pasteur was required to show, according to the original terms of the competition, was that air could be admitted to some flasks without change to their content. After failing to change the terms of reference, Pouchet withdrew, believing that he would be unable to obtain a fair hearing given the biasses of the members of the commission.

Poucher's position could not be maintained in the face of his twice withdrawing from competition. That the commissions were entirely one-sided in their views was irrelevant to a scientific community already almost uniformly behind Pasteur.

Retrospect and Prospect on the Pasteur-Pouchet Debate

Pouchet's position was rather like that of an accused person whose fate hangs on forensic evidence. Granted, the accused was given the chance of producing some evidence of his own, but the interpretation was the monopoly of the "prosecution" who also acted as judge and jury. It is easy to see why Pouchet withdrew. It is also easy to understand how readily Pasteur could claim that Pouchet's Pyrenean experiments were confounded by his use of a file rather than pincers to cut the neck of the flasks. We can imagine the fragments of glass, somehow contaminated by the file even though it had been heated, falling into the infusion of hay and seeding the nutrients therein. We can imagine that if Pouchet had been forced by the commission to use sterilised pincers after the fashion of Pasteur then many of the flasks would have remained unchanged. We may think, then, that Pouchet's understandable failure of nerve in the face of this technical straitjacket merely saved him from a greater embarrassment. Although the two commissions were disgracefully biassed, surely this was merely a historical contingency that would not have affected the accurate scientific conclusion they reached?

Interestingly, it now seems that if Pouchet had not lost his nerve he might not have lost the competition. One difference between Pouchet and Pasteur was the nutritive medium they used for their experiments, Pasteur using yeast and Pouchet hay infusions. It was not until 1876 that it was discovered that hay infusions support a spore that is not easily killed by boiling. While the boiling of a yeast infusion will destroy all life, it does not sterilise a hay infusion. Modern commentators, then, have suggested that Pouchet might have been successful if he had stayed the course—albeit for the wrong reasons! It is worth noting that nowhere do we read of Pasteur repeating Pouchet's work with hay. In fact, except to complain about the use of a file instead of pincers, he hardly ever mentioned the Pyrenean experiments, expending most of his critical energy on the earlier mercury-trough experiments for which he had a ready-made explanation. The Pyrenean experiments, of course, were carried out without mercury, the supposed contaminant in the earlier work. As one of our sources remarks: "If Pasteur ever did repeat Pouchet's experiments without mercury, he kept the results private." The conclusion to the debate was reached, then, as though the Pyrenean experiments had never taken place.

The difference between hay and yeast, as we now understand it, adds a piquant irony to the results of the commission. We, however, do not think that Pouchet would have been wiser to go ahead with the challenge, and that scientific facts speak for themselves. The modern interpretation suggests that the facts of hay infusions would have spoken, even to a biassed commission, in the unmistakable voice of spontaneous generation. We don't believe it. The commission would have found a way to explain Pouchet's results away.

Postscript

It is interesting that the defenders of Pasteur were motivated in part by what now seems another scientific heresy. It was thought at the time that Darwinism rested upon the idea of spontaneous generation. In an attack on Darwinism, published in the same year as the second commission was constituted, the secretary of the Academie des Sciences used the failure of spontaneous generation as his main argument. He wrote, "Spontaneous generation is no more. M. Pasteur has not only illuminated the question, he has resolved it!" Pasteur, then, was taken to have dealt a final blow to the theory of evolution with the same stroke as he struck down the spontaneous generation of life. One heresy destroyed another. Those who feel that because "it all came out right in the end," science is vindicated, should think again.

Finally, let us note that we now know of a number of things that might have stopped Pasteur's experiments working if he had pushed them a little further. There are various spores in addition to those found in hay that are resistant to extinction by boiling at 100°C. In the early part of the twentieth century, Henry Bastian was supporting the idea of spontaneous generation by, unknowingly, discovering more of these heat-resistant spores. Further, the dormancy of bacteria depends not only on heat but also on the acidity of the solution. Spores which appear dead in acid solution can give rise to life in an alkaline environment. Thus experiments of the type that formed the basis of this debate can be confounded in many ways. To make sure that a fluid is completely sterile it is necessary to heat it under pressure to a temperature of about 160°C, and/or subject it to a cycle of heating and cooling repeated several times at the proper intervals. As we now know, there were many ways in which Pasteur's experiments could, and should, have gone wrong. Our best guess must be that they did, but Pasteur knew what he ought to count as a result and what he ought to count as a "mistake."

Pasteur was a great scientist but what he did bore little resemblance to the ideal set out in modern texts of scientific method. It is hard to see how he would have brought about the changes in our ideas of the nature of germs if he had been constrained by the sterile model of behaviour which counts, for many, as the model of scientific method.

QUESTIONS FOR DISCUSSION

1. Early in the essay, the authors assert that "facts and reasons" are always "ambiguous." Towards the end of the essay, they say that they "do not think . . . that scientific facts speak for themselves." Why is it that scientific facts do not speak for themselves? In what way(s) are they always ambiguous?
2. The authors point out that Pasteur did not publish the results of his early experiments that seemed to confirm the spontaneous generation hypothesis. Was he right not to do so? Why would the consequences of publication have been "grave"?
3. The authors tell us that, in mid-nineteenth century France, scientific disputes were "settled" by commissions appointed by the Academy of Sciences. Are such commissions good for science?

HANS CHRISTIAN VON BAEYER

How Fermi Would Have Fixed It*

Hans Christian von Baeyer was born in Berlin into a family of doctors and scientists. After schooling in Germany, Switzerland, Canada, and New York, he earned his Ph.D. at Vanderbilt University with a dissertation on theoretical particle physics. A professor of physics since 1968 at the College of William and Mary, von Baeyer has won the two highest teaching awards of his university and in 1990 was selected as one of the outstanding faculty members of the Commonwealth of Virginia. In recognition of his contribution to mathematical physics he was elected Fellow of the American Physical Society.

In recent years von Baeyer has increasingly turned his attention to science writing for the public. His essays in The Sciences *(published by the New York Academy of Sciences),* Discover, *and other journals have won him the 1979 Science Writing Award of the American Institute of Physics, the 1990 Science Journalism Award of the American Association for the Advancement of Science, and a 1991 National Magazine Award in the category "Essays and Criticism." His books include* Rainbows, Snowflakes, and Quarks *(1984),* Taming the Atom *(1992), and* The Fermi Solution *(1993).*

In the following essay from The Sciences *(1988), von Baeyer shows that, even in the exacting world of science, "rough-and-ready" calculations have their place, as the great physicist Enrico Fermi well understood. As von Baeyer also shows, Fermi recognized, too, the importance of creativity and independence of thought in the practice of science.* *On these issues, see Bauer (this Part). For an example of independent thought at its best, see Wertheimer on Einstein (Part II).*

At twenty-nine minutes past five, on a Monday morning in July of 1945, the world's first atom bomb exploded in the desert sixty miles northwest of Alamogordo, New Mexico. Forty seconds later, the blast's shock wave reached the base camp, where scientists stood in stunned contemplation of the historic spectacle. The first person to stir was the Italian-American physicist Enrico Fermi, who was on hand to witness the culmination of a project he had helped begin.

Before the bomb detonated, Fermi had torn a sheet of notebook paper into small bits. Then, as he felt the first quiver of the shock wave spreading outward through the still air, he released the shreds above his head. They fluttered down and away from the mushroom cloud growing on the horizon, landing about two and a half yards behind him. After a brief mental calculation, Fermi announced that the bomb's energy had been equivalent to that produced by ten thousand tons of TNT. Sophisticated instruments also were at the site, and analyses of their readings of the shock wave's velocity and pressure, an exercise that took several weeks to complete, confirmed Fermi's instant estimate.

The bomb-test team was impressed, but not surprised, by this brilliant bit of scientific improvisation. Enrico Fermi's genius was known throughout the world of physics. In 1938, he had won a Nobel Prize for his work in elementary particle physics, and, four years later, in Chicago, had produced the first sustained nuclear chain reaction,

---

*Originally appeared in *The Sciences.*, Sept./Oct. 1988. Reprinted with permission of the author.

thereby ushering in the age of atomic weapons and commercial nuclear power. No other physicist of his generation, and no one since, has been at once a masterful experimentalist and a leading theoretician. In miniature, the bits of paper and the analysis of their motion exemplified this unique combination of gifts.

Like all virtuosos, Fermi had a distinctive style. His approach to physics brooked no opposition; it simply never occurred to him that he might fail to find the solution to a problem. His scientific papers and books reveal a disdain for embellishment—a preference for the most direct, rather than the most intellectually elegant, route to an answer. When he reached the limits of his cleverness, Fermi completed a task by brute force.

To illustrate this approach, imagine that a physicist must determine the volume of an irregular object—say, Earth, which is slightly pear-shaped. He might feel stymied without some kind of formula, and there are several ways he could go about getting one. He could consult a mathematician, but finding one with enough knowledge and interest to be of help is usually difficult. He could search through the mathematical literature, a time-consuming and probably fruitless exercise, because the ideal shapes that interest mathematicians often do not match those of the irregular objects found in nature. Or he could set aside his own research in order to derive the formula from basic mathematical principles, but, of course, if he had wanted to devote his time to theoretical geometry, he wouldn't have become a physicist.

Alternatively, the physicist could do what Fermi would have done—compute the volume *numerically*. Instead of relying on a formula, he could mentally divide the planet into, perhaps, a large number of tiny cubes, each with a volume easily determined by multiplying the length times the width times the height, and then add together the answers to these more tractable problems. This method yields only an approximate solution, but it is sure to produce the desired result, which is what mattered to Fermi. With the introduction, after the Second World War, of computers and, later, of pocket calculators, numerical computation has become standard procedure in physics.

The technique of dividing difficult problems into small, manageable ones applies to many problems besides those amenable to numerical computation. Fermi excelled at this rough-and-ready modus operandi, and, to pass it on to his students, he developed a type of question that has become associated with his name. A Fermi problem has a characteristic profile: Upon first hearing it, one doesn't have even the remotest notion what the answer might be. And one feels certain that too little information exists to find a solution. Yet, when the problem is broken down into subproblems, each one answerable without the help of experts or reference books, an estimate can be made, either mentally or on the back of an envelope, that comes remarkably close to the exact solution.

Suppose, for example, that one wants to determine Earth's circumference without looking it up. Everyone knows that New York and Los Angeles are separated by about three thousand miles and that the time difference between the two coasts is three hours. Three hours corresponds to one-eighth of a day, and a day is the time it takes the planet to complete one revolution, so its circumference must be eight times three thousand, or twenty-four thousand miles—an answer that differs from the true value (at the equator, 24,902.45 miles) by less than four percent. In John Milton's words:

> so easy it seemed
> Once found, which yet unfound most would have thought
> Impossible.

Fermi problems might seem to resemble the brainteasers that appear among the back pages of airline magazines and other popular publications (Given three containers that hold eight, five, and three quarts, respectively, how do you measure out a single quart?), but the two genres differ significantly. The answer to a Fermi problem, in contrast to that of a brainteaser, cannot be verified by logical deduction alone and is always approximate. (To determine precisely Earth's circumference, it is necessary that the planet actually be measured.) Then, too, solving a Fermi problem requires a knowledge of facts not mentioned in the statement of the problem. (In contrast, the decanting puzzle contains all the information necessary for its solution.)

These differences mean that Fermi problems are more closely tied to the physical world than are mathematical puzzles, which rarely have anything practical to offer physicists. By the same token, Fermi problems are reminiscent of the ordinary dilemmas that nonphysicists encounter every day of their lives. Indeed, Fermi problems, and the way they are solved, not only are essential to the practice of physics; they teach a valuable lesson in the art of living.

How many piano tuners are there in Chicago? The whimsical nature of this question, the improbability that anyone knows the answer, and the fact that Fermi posed it to his classes at the University of Chicago have elevated it to the status of legend. There is no standard solution (that's exactly the point), but anyone can make assumptions that quickly lead to an approximate answer. Here is one way: If the population of metropolitan Chicago is three million, an average family consists of four people, and one-third of all families own pianos, there are two hundred and fifty thousand pianos in the city. If each piano is tuned every ten years, there are twenty-five thousand tunings a year. If each tuner can service four pianos a day, two hundred and fifty days a year, for a total of one thousand tunings a year, there must be about twenty-five piano tuners in the city. The answer is not exact; it could be as low as ten or as high as fifty. But, as the yellow pages of the telephone directory attest, it is definitely in the ball park.

Fermi's intent was to show that although, at the outset, even the answer's order of magnitude is unknown, one can proceed on the basis of different assumptions and still arrive at estimates that fall within range of the answer. The reason is that, in any string of calculations, errors tend to cancel out one another. If someone assumes, for instance, that every sixth, rather than third, family owns a piano, he is just as likely to assume that pianos are tuned every five, not ten, years. It is as improbable that all of one's errors will be underestimates (or overestimates) as it is that all the throws in a series of coin tosses will be heads (or tails). The law of probabilities dictates that deviations from the correct assumptions will tend to compensate for one another, so the final results will converge toward the right number.

Of course, the Fermi problems that physicists face deal more often with atoms and molecules than with pianos. To answer them, one needs to commit to memory a few basic magnitudes, such as the approximate radius of a typical atom or the number of molecules in a thimbleful of water. Equipped with such facts, one can estimate, for example, the distance a car must travel before a layer of rubber about the thickness of a molecule is worn off the tread of its tires. It turns out that that much is removed with each revolution of the wheels, a reminder of the immensity of the number of atoms in a tire. (Assume that the tread is about a quarter-inch thick and that it wears off in forty thousand miles of driving. If a quarter-inch is divided by the number of revolutions a typical wheel, with its typical circumference, makes in forty thousand miles, the answer is roughly one molecular diameter.)

More momentous Fermi problems might concern energy policy (the number of solar cells required to produce a certain amount of electricity), environmental quality (the amount of acid rain caused annually by coal consumption in the United States), or the arms race. A good example from the weapons field was proposed in 1981 by David Hafemeister, a physicist at the California Polytechnic State University: For what length of time would the beam from the most powerful laser have to be focused on the skin of an incoming missile to ignite the chemical explosive in the missile's nuclear warhead? The key point is that a beam of light, no matter how well focused, spreads out like an ocean wave entering the narrow opening of a harbor, a phenomenon called diffraction broadening. The formula that describes such spreading applies to all forms of waves, including light waves, so, at a typical satellite-to-missile distance of, perhaps, seven hundred miles, a laser's energy will become considerably attenuated. With some reasonable assumptions about the temperature at which explosive materials ignite (say, a thousand degrees Fahrenheit), the diameter of the mirror that focuses the laser beam (ten feet is about right), and the maximum available power of chemical lasers (a level of a million watts has not yet been attained but is conceivable), the answer turns out to be around ten minutes.

Trying to keep a laser aimed at a speeding missile at a distance of seven hundred miles for that long is a task that greatly exceeds the capacity of existing technology. For one thing, the missile travels so rapidly that it would be impossible to keep it within range. For another, a laser beam must reflect back toward its source to verify that it is hitting its target (a process comparable to shining a flashlight at a small mirror carried by a running man at the opposite end of a football field so that the light reflected from the mirror stays in one's eyes).

The solution of this Fermi problem depends on more facts than average people, or even average physicists, have at their fingertips, but for those who do have them in mind, the calculation takes only a few minutes. And it is no less accurate for being easy to perform. So it is not surprising that Hafemeister's conclusion, which predated the President's 1983 Star Wars speech (research on laser weapons began two decades ago), agrees roughly with the findings of the American Physical Society's report entitled "Science and Technology of Directed Energy Weapons," which was the result of much more elaborate analysis. Prudent physicists—those who want to avoid false leads and dead ends—operate according to a long-standing principle: Never start a lengthy calculation until you know the range of values within which the answer is likely to fall (and, equally important, the range within which the answer is *un*likely to fall). They attack every problem as a Fermi problem, estimating the order of magnitude of the result before engaging in an investigation.

Physicists also use Fermi problems to communicate with one another. When they gather in university hallways, convention-center lobbies, or French restaurants to describe a new experiment or to discuss an unfamiliar subject, they often first survey the lay of the land, staking out, in a numerical way, the perimeter of the problem at hand. Only the timid hang back, deferring to the experts in their midst.

Those accustomed to tackling Fermi problems approach the experiment or subject as if it were their own, demonstrating their understanding by performing rough calculations. If the conversation turns to a new particle accelerator, for example, they will estimate the strength of the magnets it requires; if the subject is the structure of a novel crystal, they will calculate the spacing between its atoms. Everyone tries to arrive at the correct answer with the least effort. It is this spirit of independence, which he himself possessed in ample measure, that Fermi sought to instill by posing his unconventional problems.

Questions about atom bombs, piano tuners, automobile tires, laser weapons, particle accelerators, and crystal structure have little in common. But the manner in which they are answered is the same in every case and can be applied to questions outside the realm of physics. Whether the problem concerns cooking, automobile repair, or personal relationships, there are two basic types of responses: the fainthearted turn to authority—to reference books, bosses, expert consultants, physicians, ministers—while the independent of mind delve into that private store of common sense and limited factual knowledge that everyone carries, make reasonable assumptions, and derive their own, admittedly approximate, solutions. To be sure, it would be foolish to practice neurosurgery at home, but mundane challenges—preparing chili from scratch, replacing a water pump, resolving a family quarrel—can often be sorted out with nothing more than logic, common sense, and patience.

The resemblance of technical problems to human ones was explored in Robert M. Pirsig's 1974 book, *Zen and the Art of Motorcycle Maintenance*, in which the repair and upkeep of a machine served as a metaphor for rationality itself. At one point, the protagonist proposed to fix the slipping handlebars of a friend's new BMW motorcycle, the pride of a half-century of German mechanical craftsmanship, with a piece of an old beer can. Although the proposal happened to be technically perfect (the aluminum was thin and flexible), the cycle's owner, a musician, could not break his reliance on authority; the idea had not originated with a factory-trained mechanic, so it did not deserve serious consideration. In the same way, certain observers would have been skeptical of Fermi's analysis, carried through with the aid of a handful of confetti, of a two-billion-dollar bomb test. Such an attitude demonstrates less, perhaps, about their knowledge of the problem than about their attitude toward life. As Pirsig put it, "The real cycle you're working on is a cycle called 'yourself.'"

Ultimately, the value of dealing with the problems of science, or those of everyday life, in the way Fermi did lies in the rewards one gains for making independent discoveries and inventions. It doesn't matter whether the discovery is as momentous as the determination of the yield of an atom bomb or as insignificant as an estimate of the number of piano tuners in a midwestern city. Looking up the answer, or letting someone else find it, actually impoverishes one: it robs one of the pleasure and pride that accompany creativity and deprives one of an experience that, more than anything else in life, bolsters self-confidence. Self-confidence, in turn, is the essential prerequisite for solving Fermi problems. Thus, approaching personal dilemmas as Fermi problems can become, by a kind of chain reaction, a habit that enriches life.

QUESTIONS FOR DISCUSSION

1. In the third paragraph the author says that Fermi was at once "a masterful experimentalist and a leading theoretician." How does Fermi's measurement of the atom bomb's energy exemplify this "unique combination of gifts"?
2. What does the author mean when he says (in the fourth paragraph) that when Fermi reached the limits of his cleverness he "completed a task by brute force"?
3. What is implied (in the fifth paragraph) about mathematics (and mathematicians)?
4. What are the characteristics of the Fermi problem? What qualities and abilities was Fermi attempting to develop in his students by giving them Fermi problems? What, for instance, is the significance of the fact that "the answer to a Fermi problem . . . cannot be verified by logical deduction alone"?
5. What is the practical value for a scientist of being able to attack every problem as a Fermi problem?

DEANE RENOUF

Sensory Function in the Harbor Seal*

Deane Renouf received her Ph.D. in 1975 from Dalhousie University in Nova Scotia and is currently a professor of biopsychology at Memorial University in Newfoundland. She has done extensive research into the behavior and physiology of seals and has published in numerous professional journals. Her first book, The Behavior of the Pinnipeds, *appeared in 1991. In the following essay (from* Scientific American, *April 1989), Renouf describes the experiments she conducted to gain a better understanding of the extraordinary sensory adaptation of the harbor seal to its dark-water environment. Her description illustrates the interaction of observation, theory, experimentation, etc. involved in scientific enquiry and thus can be usefully read with Kneller's essay (this Part). Her experimental approach might well be examined in light of Popper's discussion of "falsifiability" (Part II).*

At night seals navigate through murky waters to find fish; during the day they often haul out on land, where they lie in the sun and, once a year, give birth to their young. Dividing time between land and water in such a fashion has its price: seals, like other members of the Pinnipedia (the order to which seals, sea lions, and walruses belong), have had to adapt to two separate sets of physical challenges. Sound and light behave differently in air and in water, and sensory organs that are adapted for one habitat tend to function differently in the other. Consider the seal's eye. How, if it is designed to function in water, does it also function on land? On what types of special sensory receptors does the seal rely to find its way through turbid, choppy waters?

For the past 19 years I have been studying sensory function in the harbor seal, *Phoca vitulina*. The species is a good model for understanding how pinnipeds in general have adapted to an amphibious existence and is of special interest to me because it is found in coastal waters near my home in Newfoundland. Still, after many years of observing the animal—both in its natural habitat and in captivity—I continue to be puzzled by some of its sensory capabilities.

Although it is a common species along the northern coasts of the Atlantic and Pacific oceans, *P. vitulina* is notoriously difficult to study under natural field conditions. It is extremely skittish and flees into the ocean at the slightest provocation. Such behavior makes quantitative studies nearly impossible and has discouraged many investigators. I have been lucky, however, because I have found a site on Miquelon island (some 18 kilometers off the southeast tip of Newfoundland) that is uniquely suited for observing seal colonies at close range.

During the reproductive season—in late spring and early summer—when the tide is low, about 800 seals (both male and female) cluster on exposed sand flats near the center of a large lagoon called the Grand Barachois. Females give birth at this time and remain with their pups until they are weaned at about four weeks of age. The seal's daily activities are synchronized with the tide: when the tide is high, the seals (including mothers and pups) are forced off the sand flats and into the water, where they stay until the tide recedes and the flats are again exposed. My colleagues and I have constructed elevated observation blinds adjacent to the sand flats, from which we can watch the seals at close range without disturbing them. We discovered that if we enter the blinds during high tide, the animals will pay us little or no attention when they return at low tide. From this vantage we have been able to observe how the seals have adapted to several crucial environmental challenges.

John W. Lawson, one of my graduate students, was the first to document controlled labor in the harbor seal, a physiological adaptation that allows females to accelerate or delay a pup's birth according to environmental conditions. On three occasions when a female in the final stages of labor was disturbed by the arrival of a group of tourists, Lawson saw the emerging head of the pup disappear back into the birth canal and labor come to a halt, resuming only after the disturbance had passed. We suspect that the ability of a seal to control the timing of her pup's birth is an adaptation that minimizes the risk of predation and enables seals to synchronize their labor with the onset of low tide.

While we were observing the breeding colony from our elevated blinds I became interested in the close relation that exists between females and their pups. The seal's amphibious habits and skittish behavior make bonding between mothers and their offspring somewhat problematic. I have seen a pup less than 15 minutes old follow its mother into the ocean, where visibility is often low, the current is strong and the level of ambient noise (caused by wind, choppy water and turbulence) is high. If a mother and her pup become separated, the likelihood of reunion is slim; this prediction is underscored by the fact that in some colonies as many as 10 percent of the unweaned pups starve to death every year when they are separated from their mothers.

How—if the seals rush into the ocean at the slightest disturbance—does a newborn pup manage to stay with its mother? It appears that a number of factors are responsible. The harbor seal pup demonstrates a following response (something like imprinting in birds) within the first few minutes of life and will follow its mother wherever she goes. The relationship is reciprocal: the mother in turn monitors the whereabouts of her pup. Females track their pups visually (in the water they can be seen stretching their heads backward to get an upside-down look at them) and also acoustically: a harbor seal pup vocalizes almost continuously when following its mother, emitting a call that is transmitted in air and underwater simultaneously; the call disappears from its repertoire soon after weaning.

To test whether a mother can recognize her pup by its call, Elizabeth Perry, one of my graduate students, and I recorded the calls of newborn pups and analyzed them sonographically. We found that each pup's call has a unique frequency pattern and wondered whether these differences could be discerned by a female seal.

To answer that question we devised an experiment to test a female's ability to distinguish among calls. At the Ocean Sciences Centre of Memorial University we trained a captive seal to open the door of a specific feeder when she heard one call and to open the door of another feeder when she heard a different call. Every time she made a correct association we rewarded her with herring. Six different calls were presented in various combinations; after a brief training period she was able to distinguish among the calls at least 80 percent of the time, a finding that leads us to believe females can recognize their pups in the ocean by their vocal emissions.

Vocalization is clearly an important means by which a mother and pup stay together, and yet the harbor seal lives in an environment dominated by high noise levels. How can a mother hear the call of her pup when the background noise (above water) may reach 80 decibels or more (a level comparable to that generated in an urban setting by heavy traffic)? When the ambient noise levels are high, can seals detect sounds that for human beings and other animals are masked by background noise?

I initiated a series of experiments to determine the extent to which noise affects the auditory threshold of the harbor seal. I trained two animals to swim to a paddle on one side of the tank when they heard a tone (a short burst of approximately 25 decibels, which was presented simultaneously with either 60-, 70-, or 80-decibel white noise) and to swim to a paddle on the other side if they did not hear a tone. I found that the auditory threshold of a harbor seal is raised when background noise levels are high and that the seal, like other mammals, has no special ability to compensate for noise. Extrapolating from these findings, I calculated that a pup can be heard only if it stays within about eight meters of its mother.

Within that radius of eight meters, however, females are adept at locating their young. I believe such an ability may be attributed to a unique quadraphonic hearing arrangement that enables them to determine from what direction a sound has come.

On land sound reaches the seal's inner ear—as it does in most mammals—through the auditory meatus, or canal, and its direction is determined by the difference in arrival time at each ear. (In addition, certain sound characteristics such as volume and wave pattern are affected by arrival time.) In water, where sound travels about four times faster than it does in air, the difference in arrival time is much more difficult to detect.

Bertel Möhl of the Zoological institute in Aarhus, Denmark, has shown, however, that in water sound is conducted to the seal's inner ear through a special vertical band of tissue that extends downward from the ear. When the mother's head is partially submerged, it is possible that sound passes through both the auditory meatus and the band of auditory tissue, enabling her to hear both the aerial and the underwater version of her pup's call. Because the call arrives at these receptors at slightly different times, she may be able to discern the direction from which it has come more precisely than if she relied only on the underwater or aerial versions of the call.

Harbor seals have interesting visual systems that reflect their amphibious habits. Behavioral studies by Ronald J. Schusterman and his colleagues at the University of California at Santa Cruz and anatomical studies by Glen Jamieson of the University of British Columbia and others show that the seal eye is remarkably well adapted for seeing both underwater and on land. The lens is large and spherical and its shape is suited for underwater acuity. The size and shape of the eye compensate for the fact that the refractive index of water is almost the same as that of the cornea. Consequently light waves entering a seal's eye in water do not refract, or bend, when they pass through the cornea as they do in air. Instead they are refracted only by the lens, which channels them to the retina, or focusing, plane, at the back of the eye.

In contrast, human beings, whose eyes function best on land, where the cornea is refractive, see poorly underwater. Without the help of the cornea, light is refracted by the lens so that the visual image no longer forms on the retina and the image is therefore blurred. In seals the visual image forms on the retina and is in focus.

In air the seal's cornea is astigmatic: its curvature is distorted, particularly along the horizontal plane of the eye, and light waves are affected by the distortion as they pass through the eye. In water this astigmatism is of no importance because light there is not refracted as it passes through the cornea. On land the seal compensates by having a stenopaic (vertically contracting) pupil. Because the pupil closes down to a narrow vertical slit that is parallel to the axis of least astigmatism, the most astigmatic area of the cornea has little or no effect on the seal's vision. On foggy or dimly lit beaches the pupil does not contract and the seal has blurred vision. But when light levels are higher, as they usually are near the ocean or on ice, the pupil compensates for astigmatism and the seal's visual acuity in air should be comparable to that in water.

Underwater, harbor seals are extremely sensitive to low light levels; Douglas Wartzok of Purdue University has shown, for example, that on a moonlit night in clear water the seal can detect a moving object at depths as great as 466 meters.

How does the harbor seal, which spends much of its life in murky water where visibility is near zero and which feeds mostly at night, find the three kilograms or more of fish it must catch every day? In the 1960's Thomas C. Poulter of the Stanford Research Institute and others suggested that California sea lions *(Zalophus californianus)* can find and identify prey by echolocation. Echolocation, which was first discovered in bats and was later described in birds, porpoises and dolphins, is similar in principle to radar. Animals that echolocate emit a series of high-frequency sound pulses that reflect off anything they strike; the reflections in turn are processed by the animal's brain, where they form an image that effectively enables the animal to "see" in the dark.

Although no one has conclusively demonstrated that sea lions or other pinnipeds can echolocate, a growing amount of circumstantial evidence suggests that the harbor seal may indeed have the capability. Harbor seals emit click vocalizations: broad-frequency sounds that are produced in short, very fast bursts, most often at night. Recordings I have made of these vocalizations with special audio equipment reveal that many of the clicks are in the ultrasonic range (that is, above the upper limit of human hearing, at 20 kilohertz). Working with captive seals, I have found that clicking increases when the seals are fed at night.

In 1968 Möhl was able to show that harbor seals can detect sound frequencies at least as high as 180 kHz and are most sensitive to frequencies of 32 kHz. (Human beings, in contrast, have a sensitivity range from .02 to 20 kHz.) Interestingly, some of the harbor seal clicks peaked in the range of 40 kHz, close to the seal's maximum sensitivity of 32 kHz. On land, where echolocation would be of little use to the seal (which feeds only in the water), the seal is unable to detect sound frequencies much greater than 16 kHz and emits no clicks, instead augmenting its vocabulary with low-frequency growls and snorts.

I devised an experiment with Benjamin Davis, one of my graduate students, to test echolocation in our captive seals. We wanted to see if they could distinguish between two doughnut-shaped rings that looked and weighed the same; one ring was filled with water and the other with air (and small weights), but they differed in their sound-reflecting characteristics. Because an object's density will affect the way sound waves reflect from it, we surmised that the only way the seals could differentiate between the rings was by echolocation.

We began the experiment by teaching a seal to retrieve just one of the rings at night: one of us would slip the ring into the water while the other distracted the seal at the opposite side of the training tank. The seal was then told to fetch; it quickly learned to do so, returning with the ring on its snout in an average of 34 seconds, whereupon we rewarded it with a piece of herring. Having determined that the seal could find a hidden ring and retrieve it without difficulty, we then tested the animal's ability to discriminate between the two rings. The same experimental procedure was followed except that when the seal returned with the air-filled ring, it was rewarded with a piece of herring; when it returned with the water-filled ring, a one-minute time-out (punishment for a hungry seal) was declared.

After 26 sessions the seal was able to correctly identify the air-filled ring from 75 to 80 percent of the time, a reasonable indication that it could distinguish between the two rings. We then removed the weights and added water to the air-filled ring, rendering it identical in every way with the other ring. When the experiment was repeated, the seal could no longer discriminate between them. Our results suggest that the seal can echolocate, but we are puzzled by one aspect of our study. Recordings showed that the seal vocalized very little during its search for the water-filled ring and that the clicks it emitted were intermittent and weak. Because of these findings, echolocation in the harbor seal remains unconfirmed.

Does something else explain the fact that harbor seals are extraordinarily adept at navigating and catching prey in their murky ocean habitat? For some years I have speculated that the seal's vibrissae, or whiskers, must be important sensory receptors. Vibrissae, which are present in almost all mammals (except human beings and a few other species), are unusually well developed in seals, sea lions and walruses. Research I have done with the help of my graduate student Fred Mills suggests that the vibrissae are highly sensitive to movement and thus might play a role in food capture.

We had four seals touch a small vibrating rod with their vibrissae while we varied both the frequency and the amplitude of the vibrations. By monitoring the animals' response (we gave them food when they responded to certain vibrations) we found they were most sensitive to higher frequencies (about 2.5 kHz). This finding was somewhat unexpected because it is the opposite of what occurs in other animals, whose tactile systems are most sensitive to lower frequencies. We calculated that at a distance of 43 centimeters the wave created by the tail beat of a herring-size fish would attenuate to the seal's lower threshold, theoretically enabling the seal to home in on the fish and capture it.

These predictions were given partial support when we clipped the seals' vibrissae (they grow back in a few weeks). In a set of before-and-after experiments we compared the speed at which seals could capture live fish when the vibrissae were intact with the speed at which they caught fish when the vibrissae were removed. The removal of the vibrissae had no significant effect on the length of time the seals needed to find and capture a fish in clear water. We repeated the experiment in murky water, and although they still showed no difference in prey-location time, with or without vibrissae, some dewhiskered seals took longer to actually capture fish in their mouth.

What other purpose might the vibrissae serve? In 1967 William Montagna of the Oregon Health Sciences University in Portland suggested they might function as a speedometer: their bending would correspond to the animal's swimming speed. My colleague Linda Gaborko and I set out to test this theory. We began by training a seat to swim at a speed of six kilometers per hour through hoops placed around a 17-meter course. The seal was rewarded with herring for each circuit in which it maintained a constant speed of six kilometers per hour. When it swam either too slowly or too fast, a buzzer sounded and no reward was given. Once we were convinced that the seal could maintain a steady speed (even after a 17-day break in the training), we clipped its vibrissae and repeated the trial. Its swimming speed was not affected; it thus appears that the vibrissae are important sensory receptors, but their precise function has yet to be determined.

Because the seals seem to function in many instances without conventional sensory channels, it occurred to me that they might be sensitive to magnetic fields. It is widely believed that birds can detect the earth's magnetic field and use it for compass orientation; why could the same not be true for seals? To test that hypothesis, we trained captive seals to swim through a hollow, double-walled fiberglass culvert in which we placed a Helmholtz coil: two hoops wrapped in copper wire through which a current is passed. By sending a current of 2.1 amperes at 70 volts through the coil, we were able to shift the earth's magnetic field within the culvert eastward by 60 degrees. Measurements with a magnetometer revealed that the field was uniformly shifted inside the coil and was not affected 30 centimeters outside it. Two buoys were anchored to a float three meters from the coil, one to the left of the coil, the other to the right. We hoped to train the seals to swim through the culvert on command and to touch the left buoy on emerging from the culvert when the magnetic field inside it was deflected eastward and to touch the right buoy when the coil was not powered.

The inside walls of the culvert were filled with fish oil (which prevented seawater from leaking in and also provided ballast). When the setup was complete, we conducted a number of sessions involving three seals. The first seal (a male) swam through the culvert 2,005 times in more than 30 separate sessions, the second seal (also a male) swam through it 927 times in 17 sessions and the third seal (a female) swam through it 1,227 times in 25 sessions. To our dismay the seals did not differentiate between the two electromagnetic fields.

We cannot conclude, however, that in their natural habitats seals are not sensitive to the earth's magnetic field. It must be emphasized that negative results are always difficult to interpret because they are often brought on by methodological omissions and errors. It could be that the seals failed to respond on the magnetic cues we provided because their environment (a small training tank) was artificial.

Although some aspects of seal biology are now known, many are still puzzling. It is clear that the animal, which traverses dark, murky waters in search of prey and spends considerable time on land, has either unusual sensory receptors or extraordinary cognitive abilities. Whether or not *Phoca vitulina* relies on echolocation in order to detect prey and other objects in the ocean and on the earth's magnetic field in order to navigate are questions awaiting resolution.

Questions for Discussion

1. In the first paragraph and throughout her essay, Renouf asks a number of questions. What functions do these questions serve?
2. By what means does Renouf arrive at her hypothesis that harbor seals can echolocate?
3. When describing a "genuinely empirical method," Karl Popper says in "Science: Conjectures and Refutations" (Pt II) that "confirming evidence should not count *except when it is the result of a genuine test of the theory;* and this means that it can be presented as a serious but unsuccessful attempt to falsify the theory." Describe Renouf's experiment to test echolocation in the harbor seal. Was the experiment a "serious" attempt "to falsify the theory"? Did she succeed? What does she conclude from her findings?
4. Why, according to Renouf, are negative results "always difficult to interpret"?
5. Renouf's tone throughout the essay is personal. Does this personal tone undermine the scientific objectivity of the essay?

Part II

Developing a Theory

Introduction

Just as most of us have learned that science proceeds according to a prescribed method, so it is generally believed that scientists deal only in facts and that this in turn allows them to claim the kind of certitude normally associated with facts.

But the notion that science is concerned exclusively—and unambiguously—with facts is questionable on at least two counts. In the first place, observed facts in and of themselves have no particular meaning. A flintstone with a sharp, straight edge, for instance, won't mean anything to us (or even catch our attention) unless we have a prior interest in such stones. And this interest is itself likely to be based on a more general idea (theory)—for example, about toolmaking in earlier cultures. Or again, slight physical variations among separated populations of a single species (e.g., mockingbirds) will have little interest (or significance) to an observer who doesn't already have a general understanding of the term "species" and its theoretical implications in the broad context of biological development. As the great nineteenth-century naturalist Louis Agassiz said, "Facts are stupid things until brought into connection with some general law [theory]."

In the second place, a given observation often means different things to different observers. Thus, for example, Copernicus in the sixteenth century saw the same phenomenon that Ptolemy had seen about fourteen hundred years earlier: for both, the sun rose in the east and set in the west. But whereas for Ptolemy this observation simply confirmed his preconceived notion of a geocentric universe, for Copernicus it was evidence supporting a heliocentric model. In other words, what an observer makes of observations is inevitably colored by the theory that precedes—and guides—them. In this sense, observations can be said to be theory-laden.

If the facts observed by scientists gain their meaning only in the context of a theory, which is a human construction, how then can they provide the basis for certainty? The answer, of course, is that science aims not for certitude but for understanding. We cannot know nature as it "really" is, independent of our perceptions—or of our intellectual and experimental limitations. "Natural science," Werner Heisenberg somewhere said, "is part of the interplay between nature and ourselves; it describes nature as exposed to our method of questioning." Theories are essential to that interplay—and that "method of questioning"—for they not only help guide our questions but enable us to make sense of nature's "answers" and thus lay the ground for further questions. Theories, then, are human inventions that attempt to explain the workings of nature. They must account for, and are constrained by, the "facts" but are in no way predetermined by them. They are imaginative constructions, and if (when?) nature gives them the lie, they must be discarded (or modified). This accountability to nature is the essence of scientific theory.

The essays in Part II discuss these and other aspects of theory. Charles M. Wynn in "Does Theory Ever Become Fact?" argues that, since scientific theories aim to explain every occurrence of a given observation/event, they can never claim the finality of Truth. After all, we can never be certain that Nature will not in the future offer contradictory evidence. Stephen Hawking in "My Position" examines theory from another angle. Because our understanding of "Reality" is determined by the theory that guides our perception of reality, he argues, it makes no sense (in physics, at least) to speak of reality as if it existed independently of theory. In "The Aesthetic Equation," Hans Christian von Baeyer discusses the importance of aesthetics (i.e., beauty) in theory, especially in theoretical physics. Since physicists (indeed, scientists in general) believe that the universe is a single, unified system in which the basic laws are everywhere the same, any theory or mathematical model that claims to describe this system—or any part of it—must reflect this more general belief; that is, it must be elegant, coherent, and consistent. When empirical evidence is hard to come by (as, for example, in theoretical physics), these aesthetic qualities are a sign that one is on the right track. In the next essay, Robert Oldershaw engages this very issue but emphasizes the dangers of overvaluing the aesthetic qualities of a theory. For him, such an overvaluing can lead to the ignoring of nature's evidence. In the fifth essay, "Science: Conjectures and Refutations," Karl Popper examines the essential features of scientific theories. The truly scientific theory, he insists, actively courts falsification—rather than confirmation—by making "risky" predictions. It is precisely this openness to falsification that distinguishes scientific from non-scientific theories. This distinction is taken up by Stephen Jay Gould in "Evolution as Fact and Theory." He argues that the Creationist attack on the Theory of Evolution involves a misunderstanding not only of the nature of scientific theory (and Popper's principle of falsifiability) but of science itself. The final essay of Part II focuses on the development of one of the most famous of all scientific theories. Max Wertheimer, in "Einstein: The Thinking That Led to the Theory of Relativity," dramatizes the thought processes by which Einstein arrived at his revolutionary ideas. In addition to capturing the magnitude of Einstein's intellectual and imaginative struggle, this essay illustrates the point that scientific theories are always open to modification—even those that, like Newton's, have stood fast for centuries.

CHARLES M. WYNN

Does Theory Ever Become Fact?*

Charles M. Wynn was born in New York and received his doctorate in organic chemistry from the University of Michigan. Currently a professor of chemistry and physical science at Eastern Connecticut State University, Wynn is deeply interested in science education and has published widely in this area.

*Originally appeared in *Journal of Chemical Education*, 1992, 69, 741. Reprinted with permission of the author.

In the following essay, from the Journal of Chemical Education *(1992), he illustrates the reasoning process by which scientific theories are developed and argues that such theories are by definition always tentative and therefore should never be equated with fact. Science by its very nature, he points out, is open-ended. This piece engages issues taken up most explicitly by Hawking and Gould (this Part).*

Recent breakthroughs in the field of microscopy have provided striking evidence for the apparent reality of atoms. In the 1950's, field ion microscopes were used to provide images of gas atoms striking a fluorescent screen. In 1976, scanning electron microscopes were used to produce motion pictures of the thermal motion of uranium atoms. With the development of a new family of microscopes (scanned-probe microscopes, such as the scanning tunneling microscope developed in 1981 and the atomic force microscope developed in 1985), sensing of discrete entities with properties consistent with theoretical expectations of atoms has at last been achieved. Such devices are even capable of resolving features that are only about a hundredth the size of an atom.

Now that atoms have been observed and direct evidence for the shape and location of atoms has been obtained, should the Atomic Theory be considered completely factual?

The Nature of Science

An answer to this question must be given in the context of the nature of science. Science, the study of natural and artificial phenomena in the universe with the aim of understanding them in a general way, begins with observation of phenomena. After sufficient observations have been made, a hypothesis or general statement about the basic nature of the phenomenon observed is formulated. This hypothesis should be as general as possible to deal with other phenomena besides the specific ones observed. It can thus be used to predict phenomena. Experimentation is the test of the prediction: did the phenomenon occur as anticipated? If it did, the hypothesis upon which the prediction was based gains credibility. If it did not, the hypothesis must be discarded or modified to accommodate the results.

To illustrate this approach, consider a student assigned to a room in a college dormitory. On the first day of classes, while still burrowed beneath blankets, the student can observe with his or her sense of hearing that the occupant of the adjoining room leaves the room at 6:00 A.M. and then returns out of breath at 7:00 A.M. After observing this behavior several days in a row, the student formulates a descriptive generalization of the behavior: the neighbor always leaves the dorm at 6:00 A.M. and returns at 7:00 A.M. out of breath. A hypothesis in the form of a descriptive generalization is usually called a law. In this example it might be named the Law of the Next Door Neighbor's Early Morning Behavior. One could then make a prediction based upon the law: this individual will leave the room tomorrow at 6:00 A.M. and once again return at

7:00 A.M. out of breath. If the actual behavior or experiment is consistent with the prediction, the hypothesis gains credibility. If not, it must be revised. For example, the next door neighbor's behavior may be limited to weekdays.

Once the law has been established, one could seek to explain the regularity of the behavior. A hypothesis in the form of an explanation is usually called a theory. In the example, one theory might be that the next door neighbor is a health enthusiast who behaves that way to stay fit by jogging. This might be called the Theory that the Next Door Neighbor Is a Health-Conscious Jogger. The theory predicts the observation of this neighbor jogging between 6:00 and 7:00 A.M. Until such a sighting is made, other explanations such as tennis playing, newspaper getting, etc., could also be considered.

A theory also can take the form of a model, a representation of reality invented to account for observed phenomena. Model formulation is an attempt to provide a visualization or analogue that helps one better understand the nature and interactions of the entities involved in the phenomenon. In this example, one might envision the neighbor dressed in jogger regalia—a sweatband, sweatshirt, sweatpants, and jogging shoes. This could be called the Jogger Model of the neighbor.

The same reasoning process is used by chemists. For example, after numerous determinations of mass relationships in chemical reactions, chemists formulated the laws of Conservation of Mass, Constant Composition, and Multiple Proportions. To explain these laws, they proposed the Atomic Theory, which postulates entities known as atoms.

Limited versus Universal Theories

What is the status of a Jogger Model or an Atomic Theory once the postulated entities (joggers, atoms) are actually observed?

If the next door neighbor is observed to be a jogger, the Jogger Model takes on the status of fact only for that particular jogger. While other individuals who leave their rooms at 6:00 A.M. and return at 7:00 A.M. out of breath may also be joggers, still others may be tennis players, newspaper getters, or even previously unimagined types of individuals. A theory regarding a single jogger is a limited theory, limited to a particular phenomenon. The sighting of one or even many such joggers does not mean that every early rising, out-of-breath individual is also a jogger.

Even though atomic theory has been able to explain the behavior of all matter studied thus far, and, even though in all samples scanned by microscopes, the existence of the postulated atoms has been verified, it must be considered at least conceivable that entities other than atoms might be discovered, particles which also explain the laws of Conservation of Mass, Constant Composition, and Multiple Proportions. Scientific theories can never become scientific facts, because a scientific theory deals with all instances of a phenomenon; i.e., it is a universal theory. While the behavior of all matter may indeed be explained by atomic theory, there is no way of being certain that this is the case. Such is the open-ended nature of science.

Questions for Discussion

1. "A theory," Wynn says, "can take the form of a model." What important characteristics do theory and model share?
2. According to Wynn, what is the essential characteristic of a scientific theory? Wherein lie the limits—and value—of scientific theory?
3. Wynn argues that science by its very nature is "open-ended." Does this view of science accord with those offered by Sagan (Part I)? by Gould (this Part)?

Stephen Hawking

My Position*

Educated at Oxford (B.A. 1962) and Cambridge (Ph.D. 1966), Stephen Hawking is Lucasian Professor of Mathematics at Cambridge University (a position once held by Isaac Newton) and is considered by many to be the most distinguished theoretical physicist since Einstein. His brilliance has been recognized with numerous awards, among them the Albert Einstein Medal from the Albert Einstein Society (1979), the Paul Dirac Medal and Prize from the Institute of Physics (1987), and the Wolf Foundation Prize for Physics (1988). He is a Fellow of the Royal Society and a member of the American Academy of Arts and Sciences.

As an astrophysicist, Hawking is interested in nothing less than the workings of the universe itself, including its origin and fate. Like Einstein before him, he is searching for an explanation—a "unified theory"—that will explain the behavior of everything in the universe from subatomic particles to planets and stars.

Hawking has achieved scientific preeminence despite great physical hardship. Virtually all his adult life he has suffered from ameotrophic lateral sclerosis (Lou Gehrig's disease), an affliction that has confined him to a wheelchair for many years. Since the mid-1980s, he has been unable to speak and has had to use a voice synthesizer. This handicap, however, has not hampered his ability to communicate. Travelling throughout the world, he has shared his ideas with non-specialist and specialist audiences alike. Hawking believes firmly that "the basic ideas about the origin and fate of the universe can be stated without mathematics in a form that people without a scientific education can understand." In this belief he wrote A Brief History of Time *(1988), which was on the best seller lists for more than one hundred weeks, and* Black Holes and Baby Universes *(1993), from which "My Position" is taken. Here he examines the nature and purpose of scientific theory and argues that our theories inevitably govern our perceptions of reality. In other words, in the world of physics at least, "what we regard as reality is conditioned by the*

---

*From *Black Holes and Baby Universes*, © 1993 by Stephen W. Hawking. Used by permission of Bantam Books, a division of Bantam Doubleday Dell Publishing Group, Inc. Originally given as a talk to a Caius College audience in May 1992.

theory to which we subscribe." In addition to its obvious connections with Wynn and Gould (this Part) regarding the nature and purpose of scientific theory, this essay can be usefully read with von Baeyer on the role of aesthetics in scientific theory, and with Popper on the importance of predictability/falsifiability (both this Part).

This article is not about whether I believe in God. Instead, I will discuss my approach to how one can understand the universe: what is the status and meaning of a grand unified theory, a "theory of everything." There is a real problem here. The people who ought to study and argue such questions, the philosophers, have mostly not had enough mathematical background to keep up with modern developments in theoretical physics. There is a subspecies called philosophers of science who ought to be better equipped. But many of them are failed physicists who found it too hard to invent new theories and so took to writing about the philosophy of physics instead. They are still arguing about the scientific theories of the early years of this century, like relativity and quantum mechanics. They are not in touch with the present frontier of physics.

Maybe I'm being a bit harsh on philosophers, but they have not been very kind to me. My approach has been described as naive and simpleminded. I have been variously called a nominalist, an instrumentalist, a positivist, a realist, and several other ists. The technique seems to be refutation by denigration: If you can attach a label to my approach, you don't have to say what is wrong with it. Surely everyone knows the fatal errors of all those isms.

The people who actually make the advances in theoretical physics don't think in the categories that the philosophers and historians of science subsequently invent for them. I am sure that Einstein, Heisenberg, and Dirac didn't worry about whether they were realists or instrumentalists. They were simply concerned that the existing theories didn't fit together. In theoretical physics, the search for logical self-consistency has always been more important in making advances than experimental results. Otherwise elegant and beautiful theories have been rejected because they don't agree with observation, but I don't know of any major theory that has been advanced just on the basis of experiment. The theory always came first, put forward from the desire to have an elegant and consistent mathematical model. The theory then makes predictions, which can then be tested by observation. If the observations agree with the predictions, that doesn't prove the theory; but the theory survives to make further predictions, which again are tested against observation. If the observations don't agree with the predictions, one abandons the theory.

Or rather, that is what is supposed to happen. In practice, people are very reluctant to give up a theory in which they have invested a lot of time and effort. They usually start by questioning the accuracy of the observations. If that fails, they try to modify the theory in an ad hoc manner. Eventually the theory becomes a creaking and ugly edifice. Then someone suggests a new theory, in which all the awkward observations are explained in an elegant and natural manner. An example of this was the Michelson-Morley experiment, performed in 1887, which showed that the speed of light was always the same, no matter how the source or the observer was moving. This seemed ridiculous. Surely someone moving toward the light ought to measure it traveling at a higher speed than someone moving in the same direction as the light; yet the experiment showed that both observers would measure exactly the same speed. For the next eighteen years people like Hendrik Lorentz and George Fitzgerald tried to accommodate this observation within accepted ideas of space and time. They introduced ad hoc postulates, such as proposing that objects got shorter when they moved at high speeds. The entire framework of physics became clumsy and ugly. Then in 1905 Einstein suggested a much more attractive viewpoint, in which time was not regarded as completely separate and on its own. Instead it was combined with space in a four-dimensional object called space-time. Einstein was driven to this idea not so much by the experimental results as by the desire to make two parts of the theory fit together in a consistent whole. The two parts were the laws that govern the electric and magnetic fields, and the laws that govern the motion of bodies.

I don't think Einstein, or anyone else in 1905, realized how simple and elegant the new theory of relativity was. It completely revolutionized our notions of space and time. This example illustrates well the difficulty of being a realist in the philosophy of science, for what we regard as reality is conditioned by the theory to which we subscribe. I am certain Lorentz and Fitzgerald regarded themselves as realists, interpreting the experiment on the speed of light in terms of Newtonian ideas of absolute space and absolute time. These notions of space and time seemed to correspond to common sense and reality. Yet nowadays those who are familiar with the theory of relativity, still a disturbingly small minority, have a rather different view. We ought to be telling people about the modern understanding of such basic concepts as space and time.

If what we regard as real depends on our theory, how can we make reality the basis of our philosophy? I would say that I am a realist in the sense that I think there is a universe out there waiting to be investigated and understood. I regard the solipsist position that everything is the creation of our imaginations as a waste of time. No one acts on that basis. But we cannot distinguish what is real about the universe without a theory. I therefore take the view, which has been described as simple-minded or naive, that a theory of physics is just a mathematical model that we use to describe the results of observations. A theory is a good theory if it is an elegant model, if it describes a wide class of observations, and if it predicts the results of new observations. Beyond that, it makes no sense to ask if it corresponds to reality, because we do not know what reality is independent of a theory. This view of scientific theories may make me an instrumentalist or a positivist—as I have said above, I have been called both. The person who called me a positivist went on to add that everyone knew that positivism was out of date—another case of refutation by denigration. It may indeed be out of date in that it was yesterday's intellectual fad, but the positivist position I have outlined seems the only possible one for someone who is seeking new laws, and new ways, to describe the universe. It is no good appealing to reality because we don't have a model-independent concept of reality.

In my opinion, the unspoken belief in a model-independent reality is the underlying reason for the difficulties philosophers of science have with quantum mechanics and the uncertainty principle. There is a famous thought experiment called Schrödinger's cat. A cat is placed in a sealed box. There is a gun pointing at it, and it will go off if a radioactive nucleus decays. The probability of this happening is fifty percent. (Today no one would dare propose such a thing, even purely as a thought experiment, but in Schrödinger's time they had not heard of animal liberation.)

If one opens the box, one will find the cat either dead or alive. But before the box is opened, the quantum state of the cat will be a mixture of the dead cat state with a state in which the cat is alive. This some philosophers of science find very hard to accept. The cat can't be half shot and half not-shot, they claim, any more than one can be half pregnant. Their difficulty arises because they are implicitly using a classical concept of reality in which an object has a definite single history. The whole point of quantum mechanics is that it has a different view of reality. In this view, an object has not just a single history but all possible histories. In most cases, the probability of having a particular history will cancel out with the probability of having a very slightly different history; but in certain cases, the probabilities of neighboring histories reinforce each other. It is one of these reinforced histories that we observe as the history of the object.

In the case of Schrödinger's cat, there are two histories that are reinforced. In one the cat is shot, while in the other it remains alive. In quantum theory both possibilities can exist together. But some philosophers get themselves tied in knots because they implicitly assume that the cat can have only one history.

The nature of time is another example of an area in which our theories of physics determine our concept of reality. It used to be considered obvious that time flowed on forever, regardless of what was happening; but the theory of relativity combined time with space and said that both could be warped, or distorted, by the matter and energy in the universe. So our perception of the nature of time changed from being independent of the universe to being shaped by it. It then became conceivable that time might simply not be defined before a certain point; as one goes back in time, one might come to an insurmountable barrier, a singularity, beyond which one could not go. If that were the case, it wouldn't make sense to ask who, or what, caused or created the big bang. To talk about causation or creation implicitly assumes there was a time before the big bang singularity. We have known for twenty-five years that Einstein's general theory of relativity predicts that time must have had a beginning in a singularity fifteen billion years ago. But the philosophers have not yet caught up with the idea. They are still worrying about the foundations of quantum mechanics that were laid down sixty-five years ago. They don't realize that the frontier of physics has moved on.

Even worse is the mathematical concept of imaginary time, in which Jim Hartle and I suggested the universe may not have any beginning or end. I was savagely attacked by a philosopher of science for talking about imaginary time. He said: How can a mathematical trick like imaginary time have anything to do with the real universe? I think this philosopher was confusing the technical mathematical terms real and imaginary numbers with the way that real and imaginary are used in everyday language. This just illustrates my point: How can we know what is real, independent of a theory or model with which to interpret it?

I have used examples from relativity and quantum mechanics to show the problems one faces when one tries to make sense of the universe. It doesn't really matter if you don't understand relativity and quantum mechanics, or even if these theories are incorrect. What I hope I have demonstrated is that some sort of positivist approach, in which one regards a theory as a model, is the only way to understand the universe, at least for a theoretical physicist. I am hopeful that we will find a consistent model that describes everything in the universe. If we do that, it will be a real triumph for the human race.

QUESTIONS FOR DISCUSSION

1. The author starts by asserting that his essay "is not about whether I believe in God" but about his "approach to how one can understand the universe." What, if anything, does this statement imply about Hawking's views on the relationship between God and the universe?
2. "If the observations agree with the predictions, that doesn't prove the theory. . ." (p. 64). Why don't observations that agree with a theory's predictions "prove" the theory?
3. Why are qualities such as simplicity, consistency, and elegance so important to theories in physics?
4. Explain "thought experiment." Are geologists and/or biologists more or less likely than physicists to rely on thought experiments?
5. Explain Hawking's fundamental disagreement with philosophers and historians of science. How does their need to categorize point to the nature of this disagreement?

HANS CHRISTIAN VON BAEYER

The Aesthetic Equation*

For a biographical sketch of the author, see page 46.

In the following selection from The Fermi Solution *(1993), von Baeyer discusses "beauty" in the context of science and explains why it assumes particular importance when (as in physics) experimental/observational evidence is hard to come by. On this issue of aesthetics in science, see also Hawking and Oldershaw (this Part) and Hoagland (Part III).*

Late in May 1925 the young German physicist Werner Heisenberg was staying on the island of Heligoland, in the North Sea, in an effort to rid himself of a severe case of hay fever. He was working on a radically new way to calculate the energy levels of atoms, taking into account not only the laws of classical mechanics but also the recent discovery that, in the atomic realm, energy comes in discrete bundles called quanta.

---

*Originally appeared in *The Fermi Solution*, © 1993 by Random House. Used with permission of the author.

One evening he decided to check whether his values for the energy levels obeyed the law of conservation of energy as they must in any viable theory. (The law holds that energy can be neither created nor destroyed.) Later he described what happened:

> When the first terms seemed to accord with the energy principle, I became rather excited, and I began to make countless arithmetical errors. As a result, it was almost three o'clock in the morning before the final result of my computations lay before me. The energy principle had held for all the terms, and I could no longer doubt the mathematical consistency and coherence of the kind of quantum mechanics to which my calculations pointed. . . . I had the feeling that, through the surface of atomic phenomena, I was looking at a strangely beautiful interior.

And then, too exhilarated to sleep, Heisenberg walked to the beach, climbed atop a rock and waited for the sun to rise.

Contrary to the standard picture of how science progresses, Heisenberg did not depend on experimental evidence to validate his theory; that would come later. What convinced him he was on the right track was the elegance, coherence, and inner beauty of his approach—in other words, its aesthetic qualities. Ten years earlier Albert Einstein had developed his theory of gravity in a similar manner. In general relativity, as the theory came to be called, gravity is described in terms of the geometry of the four-dimensional space-time in which we live. Einstein postulated his equations on aesthetic grounds: they were simple and consistent and, like a work of art, they felt right. "Anyone who fully comprehends this theory cannot escape its magic," Einstein wrote when he announced his creation in 1915. Experimental proof for the theory was almost entirely lacking at the time. Only one piece of evidence, a tiny anomaly in the orbit of Mercury, served as factual anchor for Einstein's bold speculation.

Today elementary-particle physicists face an experimental predicament even worse than the one Einstein did. Since the middle of the century, theorists have been guided by the data extracted from large-particle accelerators: the physical properties of the particles materialized in collisions of atomic nuclei, the directions and speeds of the emerging fragments, the changes in the processes as the energy increases, and much more. When the particles were grouped according to their masses, electric charges, and other attributes, orderly patterns appeared and pointed the way to other particles. The mathematical descriptions of these patterns led in turn to the prediction that more fundamental building blocks should exist. These too were found and named quarks. Theory and experiment inspired each other and kept pace. But eventually accelerator experiments became so time-consuming and expensive that theorists overtook their experimental colleagues.

Now theory is far ahead of experiment; indeed, it is out on a limb. The latest candidate for a unified description of all forces and matter, the so-called string theory, is so far removed from experimental testing that it has been called a theory of the twenty-first century accidentally discovered in the twentieth. The theory simplifies physics by assuming that all observed particles are different manifestations of the same fundamental entity. It also predicts, however, that this merging of identities cannot be observed except under conditions of heat and pressure that are inaccessible to current technology or, for that matter, to that of the foreseeable future. There is far less evidence for string theory today than there was for general relativity in 1915.

Proponents of the theory justify their creation by pointing to its elegance, coherence, and beauty. One of the attractions of the theory, for example, is that it combines general relativity with quantum mechanics. There is no law of nature requiring that these two theories, one for the microworld and the other for the universe at large, fit into a single mold. But the twin hopes for unity and simplicity, both aesthetic criteria, are so strong that theorists have pursued a quantum theory of gravity for more than fifty years—so far in vain. No wonder they are impressed with string theory, its remoteness from the laboratory notwithstanding. But beauty, not utility, is their guiding principle, and critics who insist that physics should not traffic with untestable propositions dismiss string theory as recreational mathematics. In reply to this calumny, string theorist David Gross of Princeton University quotes the late Nobel laureate Paul Dirac, a founder of quantum theory: "The research worker in his efforts to express the fundamental laws of nature in mathematical form should strive mainly for mathematical beauty."

None of this sounds a death knell for experiment, of course. Wherever experimental evidence can be coaxed out of nature, it suffices to corroborate or refute a theory and serves as the sole arbiter of validity. But where evidence is sparse or absent—as it is for a growing number of questions in physics—other criteria, including aesthetic ones, come into play in an essential way, both for formulating a theory and for evaluating it. In view of this fact, it is imperative that physicists know what they mean when they make appeals to such standards as elegance, coherence, and inner beauty. Many professional scientists use these terms to refer to their work, but few take the trouble to define them. What, then, is meant by elegance? by coherence? And what is beauty, in the context of mathematical formulas and physical theories?

One man who has thought deeply about these questions is Subrahmanyan Chandrasekhar, a professor emeritus at the University of Chicago. His name is as well known within the physics community as it is unknown without. His main achievement is the proof that stars with more than 1.4 times the mass of our sun collapse when they die, whereas lighter ones escape that fate. (We now know that the collapse gives rise to a neutron star or a black hole.) This work dates to 1931, when Chandrasekhar was in his twenties and beginning graduate study at the University of Cambridge. But it was not only because of one spectacular feat of mathematical reasoning that he won the Nobel Prize. Chandrasekhar was honored for a spectrum of contributions to astrophysics, ranging from a complete description of stellar and planetary atmospheres to a mathematical theory of black holes.

In an autobiographical note to his Nobel lecture, Chandrasekhar described how his career has followed a highly ordered pattern. He would carefully pick a topic that suited his tastes and abilities, and spend several years studying it in depth. Then, when he felt he had achieved a perspective of his own on the entire subject, he would summarize it systematically in a "coherent account, with order, form, and structure." In this way he produced the seven monographs that constitute his scientific oeuvre.

Throughout his life Chandrasekhar has been preoccupied with the role of aesthetics in physics. In the introduction to his first book, a treatise on the theory of stellar structure that has served as the standard in its field for half a century, he struck the chord that was to reverberate through all the rest of his work. The book begins with an exposition of thermodynamics, the science of heat. This topic had been described countless times, and another summary seemed superfluous in a text on astrophysics. But

Chandrasekhar had his own point of view. The treatment of thermodynamics he presented, a highly structured mathematical formulation, had been developed recently in Germany and was new to the English-speaking world. In Chandrasekhar's words this was, of all the treatments of the subject, "the only physically correct approach to the second law." And then he went on to remark that the logical rigor and beauty of the exposition was to exemplify the standard of perfection that should be demanded of any theory, including his own.

Chandrasekhar returned to the subject of beauty again and again, not only in his technical work but also in speeches and essays spanning forty years. In 1987 the latter were collected and published under the title *Truth and Beauty: Aesthetics and Motivations in Science*. The book is not a monolithic "coherent account," and much of it is so technical it is difficult even for physicists to read. Yet in the end it succeeds: through examples, such as the story of Heisenberg's revelation, through the accumulation of quotations from scientists and artists, and through accounts of his own experiences, Chandrasekhar conveys the perspective he has attained in a lifetime of creating theoretical physics and reflecting on its meaning.

In the pivotal essay of the collection, Chandrasekhar sets himself the difficult task of defining mathematical beauty. (He declines to take the easy way out, as Dirac, for one, did with his remark that mathematical beauty is as indefinable as artistic beauty, but is obvious when you encounter it.) As the title of Chandrasekhar's book suggests, he appreciates the power of Keats's insight that beauty is truth and truth beauty, but he wants to be more specific. Among the qualities of mathematical beauty he singles out, the most compelling is a sense of proportionality, of relatedness of parts to one another, of orderliness. According to Chandrasekhar, Heisenberg put it best: "Beauty is the proper conformity of the parts to one another and to the whole."

This definition can be applied to works of art as well as to works of mathematics. It suggests a feeling of balance, of parts that resonate rather than clash, of a total unity that bears an appropriate relation to its components. But concepts such as conformity, proportion, and order—like elegance, coherence, and even beauty itself—remain vague. What makes some things conform to one another, while other things don't? Why does one part fit naturally and beautifully into the context of the whole, while another one seems out of place?

One suggestion that renders the whole discussion more concrete can be found in the 1956 book *Science and Human Values*, by Jacob Bronowski, creator of the television series *The Ascent of Man*. According to Bronowski, the concept that underlies conformity, proportionality, orderliness, relatedness, and unity—in short, beauty—is the simple notion of likeness. "All science is the search for unity in hidden likenesses," he wrote, and, borrowing Samuel Taylor Coleridge's definition of beauty as unity in variety, "science is nothing else than the search to discover unity in the wild variety of nature. . . . Poetry, painting, the arts are the same search."

Although artists and scientists may seek the same end, they employ techniques peculiar to their disciplines. Poets, for example, apply rhythm, rhyme, assonance, and alliteration to seduce their readers into discovering likenesses, either hidden or overt, in the work itself. In prose writing, the success of the structure, which may make use of symmetry, foreshadowing, and echoes, similarly depends on the ability of the reader to notice correspondences. Painting and sculpture display correspondences in color, form, shape, and texture. In music, likenesses are found not only in the rhythmic patterns of similar sounds but also in the more subtle relations among harmonious tones.

Such stylistic correspondences acquire meaning, and beauty, to the extent that they reflect correspondences in the content of the work itself. Bronowski illustrates the point with a line from Shakespeare. When Romeo finds Juliet in the tomb, and thinks her dead, he laments, "Death that hath suckt the honey of thy breath." The rhyme of death with breath, and the sixfold repetition of the *th* sound, sometimes silent, sometimes buzzing, are the tools of the poet's craft. But the power of the line derives from its message, the comparisons of death to a bee, of Juliet to a flower—hidden likenesses between vastly disparate things.

In physics the most primitive tool for expressing likeness is the equality of two numbers. Further, one can construct pure numbers out of measured quantities by forming their ratios. The ratio of my weight in pounds to my daughter's weight, also in pounds, is the number 3.8, with no units. Likenesses between different things can then be expressed as the equality of ratios: apples can be compared with oranges. The equality of ratios is to physics what rhythm is to poetry, and balance to painting.

Many of the great discoveries in physics ultimately boil down to equalities of two ratios. When Archimedes discovered the law of the lever, for example, he found that a balance beam is in equilibrium when the ratio of the weights is equal to the inverse ratio of the lengths of the lever arms. The equation expresses a likeness between two seemingly unrelated quantities—weight and distance—and until the law was found the likeness was hidden. Two thousand years later Galileo showed that the ratio of the acceleration of a ball rolling down an incline to the acceleration of a ball in a free fall is equal to the ratio of the height of the incline to its length.

This likeness is subtler and more deeply hidden than the law discovered by Archimedes and, for that reason, it is in some sense superior. In a similar way, works of art gain in stature, and are considered to be more beautiful, as the appearances they unify are more widely varied. Thus there is greater merit in comparing death to a bee than to, say, sleep, and more poetry in the metaphor of honey for Juliet's breath than, say, wind. A scientific theory is beautiful to the extent that the phenomena it explains are unrelated—or at least seem so.

Newton's first great discovery about gravity was his demonstration that the ratio of the acceleration of a falling apple to that of the moon is equal to the inverse ratio of the squares of the distance of the two objects from the center of the earth. To appreciate the unexpectedness of this equality, think for a moment how differently the various quantities are determined. The acceleration of an apple is measured in the laboratory, with rulers and clocks. The acceleration of the moon, on the other hand, is a highly abstract concept that involves watching that distant orb glide through the night sky, timing its return from day to day, determining its distance by some trick of celestial triangulation, deriving its orbital velocity, and finally computing the change in that velocity per second. The radius of the earth is measured either by astronomical means or by painstaking land-based geodesy. None of this has any obvious connection to falling apples. That these disparate numbers, when expressed in the appropriate ratios, should result in such a simple equation is a miracle that must have astonished even

Newton. To this day the equation remains a beautiful result of theoretical physics and amounts to the first example of the unification of seemingly disparate natural forces—celestial and terrestrial gravity. In a sense, string theory is simply a continuation of the program of unification that began with Newton.

An example of likeness established in this century was Einstein's theory that the ratio of the energies of two particles of light is equal to the ratio of their frequencies. This connection is the crucial idea underlying the quantization of energy—the notion that the energy of a wave comes in discrete quanta. Einstein's postulated relation between energy and frequency was completely unwarranted from the viewpoint of classical nineteenth-century physics: the energy carried by an ocean wave, for example, depends on other factors besides frequency, such as its height; the ratio of the energies of two ocean waves is therefore not simply the ratio of their frequencies.

The frequency-energy relation proved to be a cornerstone of quantum mechanics. It was the first hint that waves (characterized by frequency) can be thought of as particles (characterized by energy), and vice versa. Einstein received the Nobel Prize primarily for this discovery, not his more famous theories of special and general relativity. The relation was experimentally verified by the American physicist Robert Millikan, and helped lead him, too, to a Nobel Prize. That simple equality of ratios turned out to be one of the most powerful, and beautiful, unifying concepts of modern physics.

Just as the techniques of poetry go beyond rhyming, the expression of likenesses in physics can be more elaborate than the mere equality of ratios. Instead of likenesses between two phenomena, there can be family relations among the collections of facts. A prime example is the periodic table, which expresses similarities and regularities between more than a hundred elements. When the elements are merely indexed by atomic weight, a list results, but when that list is arranged so that similar elements appear in vertical columns, like the days of the week on a calendar, the list becomes the periodic table. Thus shiny metals are listed below one another, like all Tuesdays, and so are the inert gases. The power of the table comes from its ability to display the likenesses: if there is no known candidate for some entry, so that a gap appears in the table, the atomic weight of the missing element can be inferred from its position in a horizontal row, and its physical properties from those of its vertical neighbors. In this way the exact properties of yet undiscovered elements have been predicted theoretically, and subsequently confirmed in the laboratory. The classification of subatomic particles and the discovery of new ones have proceeded in the same fashion. The beauty of both the periodic table and the classification of elementary particles resides in their power to expose hidden likenesses.

The search for order need not be confined to objects such as atoms and elementary particles. At the highest level of theoretical physics, it applies to formulas and equations as well. The mathematical expressions themselves become objects of contemplation. Like a collector sorting seashells, the theoretical physicist plays with equations, writing them out in different forms and combinations, looking for likenesses until an orderly and pleasing arrangement is found.

In the nineteenth century, James Clerk Maxwell noticed that the four known equations of electricity and magnetism almost—but not quite—formed a striking pattern. When he exchanged the symbols for the electric and magnetic fields, he got back almost the same equations he had started with. To satisfy his aesthetic sensibilities, Maxwell boldly modified the last formula, making the symmetry perfect. The new equations showed him a profound hidden relation: an oscillating electric field gives rise to an oscillating magnetic field, and vice versa. Maxwell realized that this reciprocity could lead to a kind of bootstrapping effect, whereby the oscillating magnetic and electric fields would mutually sustain each other. It was a discovery that immediately led him to two important applications of the theory: his prediction of the existence of radio waves and his explanation of the nature of light. What started as an attempt to impose order and consistency on a set of equations ended with the most elegant and powerful theory of classical physics, and another giant step toward the unification of all forces.

In the final essay of his book, Chandrasekhar describes one of his own theoretical discoveries in order to illustrate the role of aesthetics in mathematical creation. The context is the general theory of relativity, and the subject is black holes. Einstein's equations describe exactly how space and time are intertwined and curved in the vicinity of one of these exotic objects. The black hole itself can be thought of as a star so compact that even light cannot escape the powerful pull of its gravity; to an outside observer the star is invisible. If the black hole is perfectly round and rotating about its own axis, the shape of space around it is described by just two variables: the distance from a point in space-time to the center of the star, and the latitude on the star to which the point corresponds. Time does not play a role, because nothing changes, and neither does longitude, because the black hole is symmetric about its axis of rotation. Under such conditions general relativity prescribes one equation in two variables.

Through years of study Chandrasekhar had become thoroughly familiar with the solutions of the equations describing the neighborhood of a black hole. He had also begun to look at colliding gravitational waves, about which much less was known. (Although astronomers generally agree that black holes exist, there is not yet any observational evidence for colliding gravitational waves.) Oddly enough, the interaction of two waves crashing head-on can also be described with one equation in two variables. If the waves are thought of as water waves approaching each other from opposite directions in a long, straight canal, only their positions along the canal, and time, play a role. Both waves can be conveniently measured from the time and location of their collision. Distance from the sides of the canal or above and below the waterline is not important. Thus to describe both round, stationary black holes and colliding gravitational waves, general relativity gives one equation in two variables. The difference is that in the former case both variables are related to position, whereas in the latter case one of them is time.

To his astonishment, Chandrasekhar found that if he transformed both equations through a series of abstract and formalistic tricks, they became not just similar but identical. The identity does not imply that the two physical phenomena are identical: how could a stationary black hole resemble two colliding waves? Instead the identity is only formal; the syntax of the two equations is the same, even though the interpretations of the two sets of symbols are entirely different. And yet, a hidden likeness had come to light.

A similar likeness appears in a much simpler context. If there is no air resistance, a cannonball shot horizontally from a cliff follows a path through space that mathematicians describe as half a parabola. The two variables of the problem are horizontal distance and vertical height, both measured from the cannonball's starting point. A ball rolling down an inclined plane is also described by two variables: the distance it has moved and the time elapsed. When the distance is plotted against the time, the figure that emerges is, mirabile dictu, also half a parabola. Just as in the case of general relativity, the two phenomena are dissimilar; the relevant variables include time in one case but not in the other, and the two identical mathematical descriptions have quite different physical interpretations. But the exposed likeness points to a beautiful and profound relation between the phenomena.

Chandrasekhar exploited the similarity he discovered between black holes and colliding gravitational waves to add a multitude of new and unexpected insights to the study of the latter—"Implications," as he put it, that "one simply could not have foreseen." In his essay he emphasizes that only his sense of aesthetics, his firm belief in the beauty of general relativity, enabled this development. Without such a guide, and in the absence of experimental evidence, he would never even have imagined the transformations of the variables that made the two equations look the same. On the last five pages of *Truth and Beauty* are tables of mathematical symbols, quite impenetrable to the lay person, that describe black holes and colliding waves. But Chandrasekhar points out that the beauty of the scheme is apparent even to the untutored eye: "The pictorial pattern of this table is a visible manifestation of the structural unity of the subject."

It must be admitted that Chandrasekhar's monograph does not quite meet the standards of order, form, and structure that he set with the other, more technical chapters of his life's work. But then again, beauty, for all its importance in helping solve technical problems, is itself stubbornly resistant to codification. Perhaps the process is more important than the product, and the search for beauty more significant than its definition. If that is so, Chandrasekhar has succeeded brilliantly. By tirelessly seeking out hidden likenesses in some of the most abstract sectors of theoretical physics, Chandrasekhar has become the very embodiment of the quest for mathematical beauty in science.

QUESTIONS FOR DISCUSSION

1. What do you take Heisenberg to mean when he says that, no longer able to "doubt the mathematical consistency and coherence of the kind of quantum mechanics to which my calculations pointed. . . . I had the feeling that, through the surface of atomic phenomena, I was looking at a strangely beautiful interior." Why strange? In what sense can mathematics be said to offer a glimpse of a "beautiful interior"?

2. Would von Baeyer (and Heisenberg and Einstein) agree with Hawking (see previous essay) on the nature of the relationship between theory and experiment in theoretical physics?

3. Explain why theory is far ahead of experiment in theoretical physics. What dangers do you see in this imbalance? What are its implications regarding science/scientific method?

4. In comparing Galileo's discovery to Archimedes' (p. 71), von Baeyer says that Galileo's was "in some sense superior" because it is "subtler and more deeply hidden." Why does "subtler and more deeply hidden" amount to "superior"?
5. Explain the power of "beauty" in science. Do you see a connection between aesthetics and prediction?

ROBERT OLDERSHAW

What's Wrong with the New Physics?*

The author of some forty-five papers and articles, Robert Oldershaw is an independent researcher whose main interests lie in physics, astronomy, and cosmology. In the following piece, taken from the New Scientist *(Dec. 1990), he identifies what he takes to be two disturbing trends in contemporary theoretical physics: an overconfidence in mathematically elegant but unverifiable theories, and a reluctance to discard such theories even when observations do not support them. Such an approach, he suggests, violates a fundamental principal of science—that nature itself must be the final arbiter of a theory's value. With respect to this point, see also Hawking and Popper; with regard to "elegant theories," see von Baeyer (all three this Part).*

Recently, an astrophysicist, commenting in the highly reputable journal *Nature* on current theories of star formation, described one hypothesis as a "certainty" and another as a "near-certainty." He then made an even more remarkable statement that "there is no observational confirmation of either the certainty or the near-certainty." What has happened to the scientific method? In the first place, science does not deal in "certainty." Let's leave absolute truth to religion. In the second place, it is unscientific to put so much confidence in hypotheses that have not been confirmed by observations.

This example of scientific impropriety is hardly an isolated case. During the past decade or so, two worrying trends have emerged in the two areas of physics that claim to explain the nature of everything—particle physics and cosmology. The first trend is that physicists are increasingly devising mathematically elegant hypotheses, which they say are "compelling" but which nevertheless cannot be verified by experiments or observations. The second trend is that theorists are becoming reluctant to give up their elegant notions, preferring to modify the theory rather than discard it even when observations do not support it.

---

*Originally appeared in *New Scientist*, December 1990. Used with permission of the author.

We all learnt at school that the basic steps of scientific research are: first, to study nature; secondly, develop hypotheses about how things work or how they are correlated; thirdly, pursue the hypotheses until they yield unique and testable predictions; and last, test the truth of the hypotheses by careful observations or experiments. Of course, in reality, things are rarely that neat. Serendipity, for instance, has always been a star player in scientific progress, predictions are never perfect, and empirical results can be uncertain or misleading. But none of these factors fundamentally undermines the basic scientific method, which has rapidly increased our knowledge of nature over the past century.

As an example of how scientific method should work, consider the following. Earlier in this century, astronomers observed many spiral nebulae distributed over the celestial sphere. They devised two hypotheses to explain the spiral nebulae. The first was that they were relatively small, local clouds of gas and dust; the second was that they were separate "island universes" of stars, as proposed by Immanuel Kant, Johann Lambert, Edmund Fournier d'Albe and others. To cut a long story short, observations gradually revealed that the spiral nebulae were mainly composed of stars and were at huge distances from us. Thus humans made the momentous discovery that matter is organised into the vast island universes we now call galaxies.

The third step in the scientific method—probing new hypotheses for unique and testable predictions—is crucial to the integrity of science. Einstein's theory of relativity predicted the gravitational red shift of light, for example, which was verified by observation. Without such definitive predictions, there is no reliable way to distinguish between a true scientific insight and the merely slick. Yet increasingly, physics is presenting us with hypotheses that do not, and in some cases cannot, yield definitive predictions. Physicists would very much like to develop a unified theory for the particles of matter and the four fundamental forces of nature, the electromagnetic force, the strong and weak nuclear forces and gravity. They hope that such a framework will provide an ultimate theory of how the Universe works—a "Theory of Everything." It is a laudable aim for science and there has been real progress in this century.

The so-called Standard Model of particle physics, which describes the relationship between the fundamental particles and also unifies two of the four forces of nature, electromagnetism and the weak force into the electroweak force, has been moderately successful. It has predicted the existence of new particles, for example, the W and Z particles discovered in experiments in 1983 at CERN, the European Laboratory for Particle Physics. Other particles, the Higgs particle and the top quark, which must exist if the Standard Model is right, have not yet been detected in any of the world's accelerators.*

Nevertheless, these experimental shortcomings have not prevented theorists from charging ahead and trying to unify the electroweak force with the strong nuclear force in what are called Grand Unified Theories (GUTs). Here predictions become a problem but the particle theorists have called on the cosmologists for help. Apparently, the

*Editors' note: Good evidence for the top quark has recently been found.

most crucial aspects of grand unification happened when the Universe was 10^{-25} seconds old, just after it had exploded into existence in the "big bang." The theorists say that before this moment there was one force which then separated into the forces we know now. So far, we have no way of observing this putative blip of cosmological harmony, nor can we rerun the big bang to check.

The big bang hypothesis has theoretical problems anyway. It requires the concept of "inflation," a brief period when the early Universe is supposed to have expanded rapidly. Inflation is based on Grand Unified Theory, and suffers from the same inability to be reliably tested. Its crucial predictions are forever confined to the unobservable past while the predictions that apply to the Universe as it is now could be the result of other causes.

The most obvious area where predictability breaks down is when our intrepid theorists go on to unify all four forces of nature, that is, the electroweak force, the strong force and gravity—a so-called theory of everything. Here another "compelling" mathematical framework called string theory steps in. Instead of treating fundamental particles as dimensionless points, string theory envisages these particles as unimaginably tiny (on the order of 10^{-33} centimetres) vibrating "strings." The idea is attractive because it is mathematically satisfying and does have the potential for unifying the four fundamental forces, which has been a major theoretical problem. A lot of effort has been diverted into string theory yet it has not led to a single prediction.

In addition to these well-known theories, there are many other hypotheses of the "new physics" that suffer from a lack of testable predictions. Some that come to mind are the existence of "hidden dimensions," "shadow matter," "wormholes" in space-time and the "many worlds" interpretation of quantum mechanics. Speculation is a crucial part of scientific progress and it must be broadly encouraged. But without the benefit of predictions, we are in serious danger of ending up with elegant theories that have little or nothing to do with how the real world works.

Another troubling trend of the new physics is that the theories have many arbitrarily adjustable parameters (one model fits all data), or they come in many slightly different versions, so as to hedge one's bets. Although these theories do make predictions, their effectiveness is compromised by excessive flexibility. The strategy follows the scientific method for the first three steps, but the fourth step is changed to something like this. Test the predictions and if they are not born out experimentally, then achieve agreement, or at least avoid conflict, by twiddling with the adjustable parameters or switching to a slightly modified version of the theory.

An irreverent name for this strategy might be the Ptolemaic method. Remember Ptolemy, the Greek astronomer-mathematician who, having accepted as fact the hypothesis that the Earth was the centre of the Universe with the Sun, Moon and planets revolving round the Earth, developed a theory of planetary motion that involved adding increasingly complicated "epicycles" until his predictions fitted the facts.

The Ptolemaic method has been lavishly applied to both the Standard Model of particle physics and the big bang cosmological model. The Standard Model, believe it or not, has no less than 20 arbitrarily adjustable parameters. Moreover, various "epicycles" such as quark confinement, "charm," gluons, the Higgs mechanism, the "quark-antiquark sea," renormalisation and a hugely negative cosmological constant have been added over the years to keep it consistent with experimental results and mathematical constraints.

The big bang cosmological model has several serious problems and the inflation hypothesis, which I mentioned before, was brought in to rescue it. When the original inflation model ran into contradictions, it was replaced by a modification called the "new inflation." When further problems arose, theorists postulated yet another version called "extended inflation." Some have even advocated adding a second inflationary period—"double inflation."

A good example of the Ptolemaic strategy is the sorry case of magnetic monopoles, which are hypothetical elementary particles with a single north or south magnetic pole. These are predicted by the Grand Unified Theories that I already mentioned. For several decades, experimenters have used ingenious methods to search for these "unicorns," but to no avail. In response, theoreticians have shown an even more dazzling ingenuity in adjusting the properties of magnetic monopoles, or theoretical predictions of their hiding places, so that the legend of magnetic monopoles could be kept alive. In the latest dodge, and this one is truly unbeatable, theoretical physicists have decided that the most probable number of magnetic monopoles in this incredibly vast Universe is just one. Right. So no wonder we haven't found it yet. The little devil is probably lurking behind some quasar a billion light years away. Should we simply stop looking for it and take its "reality" on faith in our theoretical wizards?

Indeed, there is a veritable zoo of popular hypotheses of the new physics such as cold dark matter, cosmic strings, and weakly interacting massive particles (WIMPs), all needed to justify various aspects of modern cosmological theory, and each has undergone repeated adjustments or modifications in order to avoid conflict with expanding knowledge obtained from observations.

What exactly have been the major successful predictions of particle physics and cosmology in recent decades? If you discount retrodictions—"predictions" made after the observational discovery has been made—then one is hard pressed to come up with good examples. The successful prediction of the masses of the Z and W particles did show that the Standard Model at least has heuristic value. Cosmologists predicted that some radiation should have remained after the big bang—the 3 K microwave background. Astronomers found this background radiation, though the amazing homogeneity of this radiation may now force a reinterpretation of its physical origin.

But most of the fundamental discoveries in recent decades, and there have been many—have come as a complete surprise, leaving theoreticians scrambling to apply their Ptolemaic method. We are left with theories that can reproduce the data, at least approximately, but so did the original Ptolemaic model of the world.

The problem is that theory is far outpacing experimental observations. Key experiments in particle physics and astrophysics can take a decade or more to implement, from planning to publication. This slow observational "response time" permits speculative theories to become firmly entrenched in the hearts and minds of the new physicists long before the ideas can be adequately tested. This Platonic attitude worries me—a hypothesis can come to be regarded as being so convincing and elegant that it simply has to be right. This then leads researchers to mistrust and neglect observational results that conflict with the hypothesis. A classic example is surely our relentless devotion to the traditional paradigm of the big bang in cosmology.

In the beginning, so to speak, the big bang was a reasonable and thoroughly scientific interpretation of a very small amount of cosmological data. The strong evidence for very large-scale expansion, the ease with which the model could be made to account for the known abundances of different elements, such as hydrogen and helium, and the prediction of a microwave background at approximately the right temperature were quite consistent with the working hypothesis that about 15 billion years ago the cosmos was a dimensionless point—a "singular" state—and then went bang.

Paradoxically, however, as theoretical and observational problems began to pile up, the depth and breadth of the confidence in this basic paradigm increased. Let us consider some of those problems. First, the big bang is treated as an unexplainable event without a cause. Secondly, the big bang could not explain convincingly how matter got organised into lumps (galaxies and clusters of galaxies). And thirdly, it did not predict that for the Universe to be held together in the way it is, more than 90 per cent of the Universe would have to be in the form of some strange, unknown dark form of matter.

Even the strongest piece of evidence for the big bang has turned on it. Matter is not found to be spread out uniformly. Correspondingly, the leftover radiation from the big bang should also be inhomogeneous. Unfortunately, the results from the Cosmic Background Explorer (COBE) satellite, recently launched to investigate the microwave background, has revealed that this wash of radiation is relentlessly uniform. So it conflicts with the theoretical big bang predictions.

Nevertheless, the theorists are determined to hang on in there. Before COBE was launched, cosmologists predicted that fluctuations in the microwave background radiation would have to be found at more than one part per 100,000, or the big bang model would be in serious trouble. That "line in the sky" now appears to have been crossed but the response has been Platonic retrenchment: the big bang has to be right, fluctuations will show up by the one part per million level, or else.

Theorists also invented the concepts of inflation and cold dark matter to augment the big bang paradigm and keep it viable, but they too have come into increasing conflict with observations. In the light of all these problems, it is astounding that the big bang hypothesis is the only cosmological model that physicists have taken seriously.

Because experimental verification and observations have lagged behind theory, researchers have increasingly invoked theoretical "elegance" and the "potential for unification" as criteria for judging a theory's scientific merit. Beauty in science can be highly subjective; a theory that is elegant and compelling to one scientist might appear ungainly and arbitrary to another. The prediction and observation steps of the scientific method are not infallible as a means to decide theoretical questions, but they provide the best criteria that we will ever have. Without them science is in trouble.

Nevertheless, we need not be completely gloomy. While particle physics and cosmology seem to have headed off to never-never land, other sectors of physics have experienced genuine progress in recent decades. Advances with new superconducting materials have been very impressive. The application of fractal geometry to many areas of physics has led to a remarkable string of successes. Observational astrophysics has served up a veritable cornucopia of important discoveries. The new science of chaotic systems has initiated a revolution in our understanding of dynamics, and many other areas of condensed matter, nuclear, atomic and chemical physics have seen important gains.

One theme that underlies all this progress is that experimental and observational work has either been the primary driving force or has shared centre stage with theoretical work. The development of relatively high-temperature superconductors has come despite mistaken theoretical expectations, fractal geometry has remained an entirely descriptive phenomenon, and the science of chaos began, and has made subsequent advances, purely by serendipity. Science works well when observation leads theory in the dance of progress, or when the lead alternates between observation and theory on a reasonably short timescale.

How can we achieve real progress in particle physics and cosmology? Obviously, we need to emphasize the predictive powers of theories and corresponding experimental tests. Criteria such as "beauty" and the "potential for unification" can be useful in evaluating theories, but we shouldn't rely too much on such subjective judgments. We must be more honest and forthright about the purely speculative nature of some of our most cherished assumptions. Repeating them as "givens" does not make them more accurate, though it does foster that unfortunate illusion.

Some theorists have openly expressed the view that "physics is almost finished"; if these Platonists have their way physics, as a science, will indeed be finished. Theorists must accept and be guided by observations that contradict their theoretical expectations. We ignore nature's verdicts at our own peril.

QUESTIONS FOR DISCUSSION

1. For Oldershaw, the essential flaw in much of the "new physics" is that its theories so often fail to offer definitive predictions. Why is prediction crucial to scientific theory?
2. Why is "flexibility" not a virtue in scientific theories?
3. What do you understand Oldershaw to mean when he describes some of the "new physicists" as Platonists? What broader point about contemporary physics might he be making here?
4. How would you describe the author's tone (i.e., his attitude towards the new physics/physicists)? Is it appropriate to scientific discourse?

KARL POPPER

Science: Conjectures and Refutations*

Karl Popper, Austrian philosopher of natural and social science, was born in Vienna and studied mathematics, physics, and philosophy at the University of Vienna. After teaching for several years at Canterbury University College in Christchurch, New Zealand, he was appointed Professor of Logic and Scientific Method in 1949 at the London School of Economics in the University of London. Popper was knighted in 1964 and until his death in 1994 remained a major intellectual force. His books include The Poverty of Historicism *(1957),* The Logic of Scientific Discovery *(1959), and* Conjectures and Refutations *(1962), from which the following essay is excerpted.*

This selection illustrates two important aspects of Popper's contributions to modern views of scientific method: 1) his ideas about "falsification" and the testing of scientific theories and 2) his use of the principle of falsification to distinguish science (and the scientific hypothesis) from non-science (from, for example, Marxist economics and Freudian psychoanalytic theory). In addition to his obvious relevance to the other essays in this Part, most especially Hawking, Oldershaw, and Gould, Popper can be usefully read with Kneller and Renouf (Part I).

I

When I received the list of participants in this course and realized that I had been asked to speak to philosophical colleagues I thought, after some hesitation and consultation, that you would probably prefer me to speak about those problems which interest me most, and about those developments with which I am most intimately acquainted. I therefore decided to do what I have never done before: to give you a report on my own work in the philosophy of science, since the autumn of 1919 when I first began to grapple with the problem, *"When should a theory be ranked as scientific?"* or *"Is there a criterion for the scientific character or status of a theory?"*

The problem which troubled me at the time was neither, "When is a theory true?" nor, "When is a theory acceptable?" My problem was different. I *wished to distinguish between science and pseudo-science;* knowing very well that science often errs, and that pseudo-science may happen to stumble on the truth.

*From *Conjectures and Refutations*, © 1962 by Basic Books. Used with permission of the executor of the estate of the author. Originally given in a lecture at Peterhouse, Cambridge, in Summer 1953, as part of a course on developments and trends in contemporary British philosophy, organized by the British Council; originally published under the title, "Philosophy of Science: a Personal Report" in *British Philosophy in Mid-Century*, ed. C.A. Mace, 1957.

I knew, of course, the most widely accepted answer to my problem: that science is distinguished from pseudo-science—or from 'metaphysics'—by its *empirical method,* which is essentially *inductive,* proceeding from observation or experiment. But this did not satisfy me. On the contrary, I often formulated my problem as one of distinguishing between a genuinely empirical method and a non-empirical or even a pseudo-empirical method—that is to say, a method which, although it appeals to observation and experiment, nevertheless does not come up to scientific standards. The latter method may be exemplified by astrology, with its stupendous mass of empirical evidence based on observation—on horoscopes and on biographies.

But as it was not the example of astrology which led me to my problem I should perhaps briefly describe the atmosphere in which my problem arose and the examples by which it was stimulated. After the collapse of the Austrian Empire there had been a revolution in Austria: the air was full of revolutionary slogans and ideas, and new and often wild theories. Among the theories which interested me Einstein's theory of relativity was no doubt by far the most important. Three others were Marx's theory of history, Freud's psychoanalysis, and Alfred Adler's so-called "individual psychology."

There was a lot of popular nonsense talked about these theories, and especially about relativity (as still happens even today), but I was fortunate in those who introduced me to the study of this theory. We all—the small circle of students to which I belonged—were thrilled with the result of Eddington's eclipse observations which in 1919 brought the first important confirmation of Einstein's theory of gravitation. It was a great experience for us, and one which had a lasting influence on my intellectual development.

The three other theories I have mentioned were also widely discussed among students at that time. I myself happened to come into personal contact with Alfred Adler, and even to co-operate with him in his social work among the children and young people in the working-class districts of Vienna where he had established social guidance clinics.

It was during the summer of 1919 that I began to feel more and more dissatisfied with these three theories—the Marxist theory of history, psychoanalysis, and individual psychology; and I began to feel dubious about their claims to scientific status. My problem perhaps first took the simple form, "What is wrong with Marxism, psycho-analysis, and individual psychology? Why are they so different from physical theories, from Newton's theory, and especially from the theory of relativity?"

To make this contrast clear I should explain that few of us at the time would have said that we believed in the *truth* of Einstein's theory of gravitation. This shows that it was not my doubting the *truth* of those other three theories which bothered me, but something else. Yet neither was it that I merely felt mathematical physics to be more *exact* than the sociological or psychological type of theory. Thus what worried me was neither the problem of truth, at that stage at least, nor the problem of exactness or measurability. It was rather that I felt that these other three theories, though posing as sciences, had in fact more in common with primitive myths than with science; that they resembled astrology rather than astronomy.

I found that those of my friends who were admirers of Marx, Freud, and Adler, were impressed by a number of points common to these theories, and especially by their apparent *explanatory power*. These theories appeared to be able to explain practically everything that happened within the fields to which they referred. The study of any of them seemed to have the effect of an intellectual conversion or revelation, opening your eyes to a new truth hidden from those not yet initiated. Once your eyes were thus opened you saw confirming instances everywhere: the world was full of *verifications* of the theory. Whatever happened always confirmed it. Thus its truth appeared manifest; and unbelievers were clearly people who did not want to see the manifest truth; who refused to see it, either because it was against their class interest, or because of their repressions which were still "un-analysed" and crying aloud for treatment.

The most characteristic element in this situation seemed to me the incessant stream of confirmations, of observations which "verified" the theories in question; and this point was constantly emphasized by their adherents. A Marxist could not open a newspaper without finding on every page confirming evidence for his interpretation of history; not only in the news, but also in its presentation—which revealed the class bias of the paper—and especially of course in what the paper did *not* say. The Freudian analysts emphasized that their theories were constantly verified by their "clinical observations." As for Adler, I was much impressed by a personal experience. Once, in 1919, I reported to him a case which to me did not seem particularly Adlerian, but which he found no difficulty in analysing in terms of his theory of inferiority feelings, although he had not even seen the child. Slightly shocked, I asked him how he could be so sure. "Because of my thousandfold experience," he replied; whereupon I could not help saying: "And with this new case, I suppose, your experience has become thousand-and-one-fold."

What I had in mind was that his previous observations may not have been much sounder than this new one; that each in its turn had been interpreted in the light of "previous experience," and at the same time counted as additional confirmation. What, I asked myself, did it confirm? No more than that a case could be interpreted in the light of the theory. But this meant very little, I reflected, since every conceivable case could be interpreted in the light of Adler's theory, or equally of Freud's. I may illustrate this by two very different examples of human behaviour: that of a man who pushes a child into the water with the intention of drowning it; and that of a man who sacrifices his life in an attempt to save the child. Each of these two cases can be explained with equal ease in Freudian and in Adlerian terms. According to Freud the first man suffered from repression (say, of some component of his Oedipus complex), while the second man had achieved sublimation. According to Adler the first man suffered from feelings of inferiority (producing perhaps the need to prove to himself that he dared to commit some crime), and so did the second man (whose need was to prove to himself that he dared to rescue the child). I could not think of any human behaviour which could not be interpreted in terms of either theory. It was precisely this fact—that they always fitted, that they were always confirmed—which in the eyes of their admirers constituted the strongest argument in favour of these theories. It began to dawn on me that this apparent strength was in fact their weakness.

With Einstein's theory the situation was strikingly different. Take one typical instance—Einstein's prediction, just then confirmed by the findings of Eddington's expedition. Einstein's gravitational theory had led to the result that light must be attracted by heavy bodies (such as the sun), precisely as material bodies were attracted. As a consequence it could be calculated that light from a distant fixed star whose apparent position was close to the sun would reach the earth from such a direction that the star would seem to be slightly shifted away from the sun; or, in other words, that stars close to the sun would look as if they had moved a little away from the sun, and from one another. This is a thing which cannot normally be observed since such stars are rendered invisible in daytime by the sun's overwhelming brightness; but during an eclipse it is possible to take photographs of them. If the same constellation is photographed at night one can measure the distances on the two photographs, and check the predicted effect.

Now the impressive thing about this case is the *risk* involved in a prediction of this kind. If observation shows that the predicted effect is definitely absent, then the theory is simply refuted. The theory is *incompatible with certain possible results of observation*—in fact with results which everybody before Einstein would have expected. This is quite different from the situation I have previously described, when it turned out that the theories in question were compatible with the most divergent human behaviour, so that it was practically impossible to describe any human behaviour that might not be claimed to be a verification of these theories.

These considerations led me in the winter of 1919–20 to conclusions which I may now reformulate as follows:

1. It is easy to obtain confirmations, or verifications, for nearly every theory—if we look for confirmations.
2. Confirmations should count only if they are the result of *risky predictions;* that is to say, if, unenlightened by the theory in question, we should have expected an event which was incompatible with the theory—an event which would have refuted the theory.
3. Every "good" scientific theory is a prohibition: it forbids certain things to happen. The more a theory forbids, the better it is.
4. A theory which is not refutable by any conceivable event is nonscientific. Irrefutability is not a virtue of a theory (as people often think) but a vice.
5. Every genuine *test* of a theory is an attempt to falsify it, or to refute it. Testability is falsifiability; but there are degrees of testability: some theories are more testable, more exposed to refutation, than others; they take, as it were, greater risks.
6. Confirming evidence should not count *except when it is the result of a genuine test of the theory;* and this means that it can be presented as a serious but unsuccessful attempt to falsify the theory. (I now speak in such cases of "corroborating evidence.")

7. Some genuinely testable theories, when found to be false, are still upheld by their admirers—for example by introducing *ad hoc* some auxiliary assumption, or by re-interpreting the theory *ad hoc* in such a way that it escapes refutation. Such a procedure is always possible, but it rescues the theory from refutation only at the price of destroying, or at least lowering, its scientific status. (I later described such a rescuing operation as a *"conventionalist twist"* or a *"conventionalist stratagem."*)

One can sum up all this by saying that *the criterion of the scientific status of a theory is its falsifiability, or refutability, or testability.*

II

I may perhaps exemplify this with the help of the various theories so far mentioned. Einstein's theory of gravitation clearly satisfied the criterion of falsifiability. Even if our measuring instruments at the time did not allow us to pronounce on the results of the tests with complete assurance, there was clearly a possibility of refuting the theory.

Astrology did not pass the test. Astrologers were greatly impressed, and misled, by what they believed to be confirming evidence—so much so that they were quite unimpressed by any unfavourable evidence. Moreover, by making their interpretations and prophecies sufficiently vague they were able to explain away anything that might have been a refutation of the theory had the theory and the prophecies been more precise. In order to escape falsification they destroyed the testability of their theory. It is a typical soothsayer's trick to predict things so vaguely that the predictions can hardly fail: that they become irrefutable.

The Marxist theory of history, in spite of the serious efforts of some of its founders and followers, ultimately adopted this soothsaying practice. In some of its earlier formulations (for example in Marx's analysis of the character of the "coming social revolution") their predictions were testable, and in fact falsified. Yet instead of accepting the refutations the followers of Marx re-interpreted both the theory and the evidence in order to make them agree. In this way they rescued the theory from refutation; but they did so at the price of adopting a device which made it irrefutable. They thus gave a "conventionalist twist" to the theory; and by this stratagem they destroyed its much advertised claim to scientific status.

The two psycho-analytic theories were in a different class. They were simply non-testable, irrefutable. There was no conceivable human behaviour which could contradict them. This does not mean that Freud and Adler were not seeing certain things correctly: I personally do not doubt that much of what they say is of considerable importance, and may well play its part one day in a psychological science which is testable. But it does mean that those "clinical observations" which analysts naively believe confirm their theory cannot do this any more than the daily confirmations which astrologers find in their practice. And as for Freud's epic of the Ego, the Super-ego, and the Id, no substantially stronger claim to scientific status can be made for it than for

Homer's collected stories from Olympus. These theories describe some facts, but in the manner of myths. They contain most interesting psychological suggestions, but not in a testable form.

At the same time I realized that such myths may be developed, and become testable; that historically speaking all—or very nearly all—scientific theories originate from myths, and that a myth may contain important anticipations of scientific theories. Examples are Empedocles' theory of evolution by trial and error, or Parmenides' myth of the unchanging block universe in which nothing ever happens and which, if we add another dimension, becomes Einstein's block universe (in which, too, nothing ever happens, since everything is, four-dimensionally speaking, determined and laid down from the beginning). I thus felt that if a theory is found to be non-scientific, or "metaphysical" (as we might say), it is not thereby found to be unimportant, or insignificant, or "meaningless," or "nonsensical." But it cannot claim to be backed by empirical evidence in the scientific sense—although it may easily be, in some genetic sense, the "result of observation."

(There were a great many other theories of this pre-scientific or pseudo-scientific character, some of them, unfortunately, as influential as the Marxist interpretation of history; for example, the racialist interpretation of history—another of those impressive and all-explanatory theories which act upon weak minds like revelations.)

Thus the problem which I tried to solve by proposing the criterion of falsifiability was neither a problem of meaningfulness or significance, nor a problem of truth or acceptability. It was the problem of drawing a line (as well as this can be done) between the statements, or systems of statements, of the empirical sciences, and all other statements—whether they are of a religious or of a metaphysical character, or simply pseudo-scientific. Years later—it must have been in 1928 or 1929—I called this first problem of mine the *"problem of demarcation."* The criterion of falsifiability is a solution to this problem of demarcation, for it says that statements or systems of statements, in order to be ranked as scientific, must be capable of conflicting with possible, or conceivable, observations.

QUESTIONS FOR DISCUSSION

1. What role does prediction play in scientific theories? What does Popper mean by a "risky prediction"?
2. In his list of conclusions, Popper says that "every 'good' scientific theory is a prohibition—it forbids certain things to happen." Why is this idea of prohibition essential to the "good" theory?
3. Why, for Popper, are attempts to falsify a theory more important than attempts to confirm it?
4. What do you understand Popper to mean by "pseudo-science"?
5. What might Popper say about theories that have out-run experiment?

STEPHEN JAY GOULD

Evolution as Fact and Theory*

Stephen Jay Gould received his doctorate from Columbia University in 1967 and is currently Professor of Zoology and curator of paleontology at Harvard University. The author of many scientific works, Gould has also written extensively for a wider audience. Books such as The Panda's Thumb *(1980),* The Mismeasure of Man *(1981), and* Eight Little Piggies: Reflections in Natural History *(1993) are much admired for their style and their accessibility to the general reader.*

In the following selection from Hen's Teeth and Horses Toes *(1983), Gould discusses Darwin's theory of evolution in the context of the Creationist attack. Central to this attack, he points out, is a misunderstanding of the nature of scientific theory and of the principle of falsifiability. These issues are engaged elsewhere in this Part; see especially Wynn, Hawking, and Popper.*

Kirtley Mather, who died last year at age ninety, was a pillar of both science and Christian religion in America and one of my dearest friends. The difference of a half-century in our ages evaporated before our common interests. The most curious thing we shared was a battle we each fought at the same age. For Kirtley had gone to Tennessee with Clarence Darrow to testify for evolution at the Scopes trial of 1925. When I think that we are enmeshed again in the same struggle for one of the best documented, most compelling and exciting concepts in all of science, I don't know whether to laugh or cry.

According to idealized principles of scientific discourse, the arousal of dormant issues should reflect fresh data that give renewed life to abandoned notions. Those outside the current debate may therefore be excused for suspecting that creationists have come up with something new, or that evolutionists have generated some serious internal trouble. But nothing has changed; the creationists have presented not a single new fact or argument. Darrow and Bryan were at least more entertaining than we lesser antagonists today. The rise of creationism is politics, pure and simple; it represents one issue (and by no means the major concern) of the resurgent evangelical right. Arguments that seemed kooky just a decade ago have reentered the mainstream.

The basic attack of modern creationists falls apart on two general counts before we even reach the supposed factual details of their assault against evolution. First, they play upon a vernacular misunderstanding of the word "theory" to convey the false impression that we evolutionists are covering up the rotten core of our edifice. Second, they misuse a popular philosophy of science to argue that they are behaving scientifically in attacking evolution. Yet the same philosophy demonstrates that their own belief is not science, and that "scientific creationism" is a meaningless and self-contradictory phrase, an example of what Orwell called "newspeak."

*Originally appeared in *Hen's Teeth and Horse's Toes: Further Reflections in Natural History by Stephen Jay Gould.*

In the American vernacular, "theory" often means "imperfect fact"—part of a hierarchy of confidence running downhill from fact to theory to hypothesis to guess. Thus, creationists can (and do) argue: evolution is "only" a theory, and intense debate now rages about many aspects of the theory. If evolution is less than a fact, and scientists can't even make up their minds about the theory, then what confidence can we have in it? Indeed, President Reagan echoed this argument before an evangelical group in Dallas when he said (in what I devoutly hope was campaign rhetoric): "Well, it is a theory. It is a scientific theory only, and it has in recent years been challenged in the world of science—that is, not believed in the scientific community to be as infallible as it once was."

Well, evolution *is* a theory. It is also a fact. And facts and theories are different things, not rungs in a hierarchy of increasing certainty. Facts are the world's data. Theories are structures of ideas that explain and interpret facts. Facts do not go away while scientists debate rival theories for explaining them. Einstein's theory of gravitation replaced Newton's, but apples did not suspend themselves in mid-air pending the outcome. And human beings evolved from apelike ancestors whether they did so by Darwin's proposed mechanism or by some other, yet to be discovered.

Moreover, "fact" does not mean "absolute certainty." The final proofs of logic and mathematics flow deductively from stated premises and achieve certainty only because they are *not* about the empirical world. Evolutionists make no claim for perpetual truth, though creationists often do (and then attack us for a style of argument that they themselves favor). In science, "fact" can only mean "confirmed to such a degree that it would be perverse to withhold provisional assent." I suppose that apples might start to rise tomorrow, but the possibility does not merit equal time in physics classrooms.

Evolutionists have been clear about this distinction between fact and theory from the very beginning, if only because we have always acknowledged how far we are from completely understanding the mechanisms (theory) by which evolution (fact) occurred. Darwin continually emphasized the difference between his two great and separate accomplishments: establishing the fact of evolution, and proposing a theory—natural selection—to explain the mechanism of evolution. He wrote in *The Descent of Man:* "I had two distinct objects in view; firstly, to show that species had not been separately created, and secondly, that natural selection had been the chief agent of change . . . Hence if I have erred in . . . having exaggerated its [natural selection's] power . . . I have at least, as I hope, done good service in aiding to overthrow the dogma of separate creations."

Thus Darwin acknowledged the provisional nature of natural selection while affirming the fact of evolution. The fruitful theoretical debate that Darwin initiated has never ceased. From the 1940s through the 1960s, Darwin's own theory of natural selection did achieve a temporary hegemony that it never enjoyed in his lifetime. But renewed debate characterizes our decade, and, while no biologist questions the importance of natural selection, many now doubt its ubiquity. In particular, many evolutionists argue that substantial amounts of genetic change may not be subject to natural selection and may spread through populations at random. Others are challenging Darwin's linking of natural selection with gradual, imperceptible change through all intermediary degrees; they are arguing that most evolutionary events may occur far more rapidly than Darwin envisioned.

Scientists regard debates on fundamental issues of theory as a sign of intellectual health and a source of excitement. Science is—and how else can I say it?—most fun when it plays with interesting ideas, examines their implications, and recognizes that old information may be explained in surprisingly new ways. Evolutionary theory is now enjoying this uncommon vigor. Yet amidst all this turmoil no biologist has been led to doubt the fact that evolution occurred; we are debating *how* it happened. We are all trying to explain the same thing: the tree of evolutionary descent linking all organisms by ties of genealogy. Creationists pervert and caricature this debate by conveniently neglecting the common conviction that underlies it, and by falsely suggesting that we now doubt the very phenomenon we are struggling to understand.

Secondly, creationists claim that "the dogma of separate creations," as Darwin characterized it a century ago, is a scientific theory meriting equal time with evolution in high school biology curricula. But a popular viewpoint among philosophers of science belies this creationist argument. Philosopher Karl Popper has argued for decades that the primary criterion of science is the falsifiability of its theories. We can never prove absolutely, but we can falsify. A set of ideas that cannot, in principle, be falsified is not science.

The entire creationist program includes little more than a rhetorical attempt to falsify evolution by presenting supposed contradictions among its supporters. Their brand of creationism, they claim, is "scientific" because it follows the Popperian model in trying to demolish evolution. Yet Popper's argument must apply in both directions. One does not become a scientist by the simple act of trying to falsify a rival and truly scientific system; one has to present an alternative system that also meets Popper's criterion—it too must be falsifiable in principle.

"Scientific creationism" is a self-contradictory, nonsense phrase precisely because it cannot be falsified. I can envision observations and experiments that would disprove any evolutionary theory I know, but I cannot imagine what potential data could lead creationists to abandon their beliefs. Unbeatable systems are dogma, not science. Lest I seem harsh or rhetorical, I quote creationism's leading intellectual, Duane Gish, Ph.D., from his book, *Evolution? The Fossils Say No!* "By creation we mean the bringing into being by a supernatural Creator of the basic kinds of plants and animals by the process of sudden, or fiat, creation. We do not know how the Creator created, what processes He used, *for He used processes which are not now operating anywhere in the natural universe* [Gish's italics]. This is why we refer to creation as special creation. We cannot discover by scientific investigations anything about the creative processes used by the Creator." Pray tell, Dr. Gish, in the light of your last sentence, what then is "scientific" creationism?

Our confidence that evolution occurred centers upon three general arguments. First, we have abundant, direct, observational evidence of evolution in action, from both field and laboratory. This evidence ranges from countless experiments on change in nearly everything about fruit flies subjected to artificial selection in the laboratory to the famous populations of British moths that became black when industrial soot darkened the trees upon which the moths rest. (Moths gain protection from sharp-sighted bird predators by blending into the background.) Creationists do not deny these observations; how could they? Creationists have tightened their act. They now argue that God only created "basic kinds," and allowed for limited evolutionary meandering within them. Thus toy poodles and Great Danes come from the dog kind and moths can change color, but nature cannot convert a dog to a cat or a monkey to a man.

The second and third arguments for evolution—the case for major changes—do not involve direct observation of evolution in action. They rest upon inference, but are no less secure for that reason. Major evolutionary change requires too much time for direct observation on the scale of recorded human history. All historical sciences rest upon inference, and evolution is no different from geology, cosmology, or human history in this respect. In principle, we cannot observe processes that operated in the past. We must infer them from results that still surround us: living and fossil organisms for evolution, documents and artifacts for human history, strata and topography for geology.

The second argument—that the imperfection of nature reveals evolution—strikes many people as ironic, for they feel that evolution should be most elegantly displayed in the nearly perfect adaptation expressed by some organisms—the camber of a gull's wing, or butterflies that cannot be seen in ground litter because they mimic leaves so precisely. But perfection could be imposed by a wise creator or evolved by natural selection. Perfection covers the tracks of past history. And past history—the evidence of descent—is the mark of evolution.

Evolution lies exposed in the *imperfections* that record a history of descent. Why should a rat run, a bat fly, a porpoise swim, and I type this essay with structures built of the same bones unless we all inherited them from a common ancestor? An engineer, starting from scratch, could design better limbs in each case. Why should all the large native mammals of Australia be marsupials, unless they descended from a common ancestor isolated on this island continent? Marsupials are not "better," or ideally suited for Australia; many have been wiped out by placental mammals imported by man from other continents. This principle of imperfection extends to all historical sciences. When we recognize the etymology of September, October, November, and December (seventh, eighth, ninth, and tenth), we know that the year once started in March, or that two additional months must have been added to an original calendar of ten months.

The third argument is more direct: transitions are often found in the fossil record. Preserved transitions are not common—and should not be, according to our understanding of evolution—but they are not entirely wanting, as creationists often claim. The lower jaw of reptiles contains several bones, that of mammals only one. The non-mammalian jawbones are reduced, step by step, in mammalian ancestors until they become tiny nubbins located at the back of the jaw. The "hammer" and "anvil" bones of the mammalian ear are descendants of these nubbins. How could such a transition be accomplished? the creationists ask. Surely a bone is either entirely in the jaw or in the ear. Yet paleontologists have discovered two transitional lineages of therapsids (the so-called mammal-like reptiles) with a double jaw joint—one composed of the old quadrate and articular bones (soon to become the hammer and anvil), the other of the squamosal and dentary bones (as in modern mammals). For that matter, what better transitional form could we expect to find than the oldest human, *Australopithecus afarensis*, with its apelike palate, its human upright stance, and a cranial capacity larger than any ape's of the same body size but a full 1,000 cubic centimeters below ours? If God made each of the half-dozen human species discovered in ancient rocks, why did he create in an unbroken temporal sequence of progressively more modern features—increasing cranial capacity, reduced face and teeth, larger body size? Did he create to mimic evolution and test our faith thereby?

Faced with these facts of evolution and the philosophical bankruptcy of their own position, creationists rely upon distortion and innuendo to buttress their rhetorical claim. If I sound sharp or bitter, indeed I am—for I have become a major target of these practices.

I count myself among the evolutionists who argue for a jerky, or episodic, rather than a smoothly gradual, pace of change. In 1972 my colleague Niles Eldredge and I developed the theory of punctuated equilibrium. We argued that two outstanding facts of the fossil record—geologically "sudden" origin of new species and failure to change thereafter (stasis)—reflect the predictions of evolutionary theory, not the imperfections of the fossil record. In most theories, small isolated populations are the source of new species, and the process of speciation takes thousands or tens of thousands of years. This amount of time, so long when measured against our lives, is a geological microsecond. It represents much less than 1 per cent of the average lifespan for a fossil invertebrate species—more than ten million years. Large, widespread, and well-established species, on the other hand, are not expected to change very much. We believe that the inertia of large populations explains the stasis of most fossil species over millions of years.

We proposed the theory of punctuated equilibrium largely to provide a different explanation for pervasive trends in the fossil record. Trends, we argued, cannot be attributed to gradual transformation within lineages, but must arise from the differential success of certain kinds of species. A trend, we argued, is more like climbing a flight of stairs (punctuations and stasis) than rolling up an inclined plane.

Since we proposed punctuated equilibria to explain trends, it is infuriating to be quoted again and again by creationists—whether through design or stupidity, I do not know—as admitting that the fossil record includes no transitional forms. Transitional forms are generally lacking at the species level, but they are abundant between larger groups. Yet a pamphlet entitled "Harvard Scientists Agree Evolution Is a Hoax" states: "The facts of punctuated equilibrium which Gould and Eldredge . . . are forcing Darwinists to swallow fit the picture that Bryan insisted on, and which God has revealed to us in the Bible."

Continuing the distortion, several creationists have equated the theory of punctuated equilibrium with a caricature of the beliefs of Richard Goldschmidt, a great early geneticist. Goldschmidt argued, in a famous book published in 1940, that new groups can arise all at once through major mutations. He referred to these suddenly transformed creatures as "hopeful monsters." (I am attracted to some aspects of the non-caricatured version, but Goldschmidt's theory still has nothing to do with punctuated equilibrium. Creationist Luther Sunderland talks of the "punctuated equilibrium hopeful monster theory" and tells his hopeful readers that "it amounts to tacit admission that anti-evolutionists are correct in asserting there is no fossil evidence supporting the theory that all life is connected to a common ancestor." Duane Gish writes, "According to Goldschmidt, and now apparently according to Gould, a reptile laid an egg from which the first bird, feathers and all, was produced." Any evolutionist who believed such nonsense would rightly be laughed off the intellectual stage; yet the only theory that could ever envision such a scenario for the origin of birds is creationism—with God acting in the egg.

I am both angry at and amused by the creationists; but mostly I am deeply sad. Sad for many reasons. Sad because so many people who respond to creationist appeals are troubled for the right reason, but venting their anger at the wrong target. It is true that scientists have often been dogmatic and elitist. It is true that we have often allowed the white-coated, advertising image to represent us—"Scientists say that Brand X cures bunions ten times faster than . . ." We have not fought it adequately because we derive benefits from appearing as a new priesthood. It is also true that faceless and bureaucratic state power intrudes more and more into our lives and removes choices that should belong to individuals and communities. I can understand that school curricula, imposed from above and without local input, might be seen as one more insult on all these grounds. But the culprit is not, and cannot be, evolution or any other fact of the natural world. Identify and fight your legitimate enemies by all means, but we are not among them.

I am sad because the practical result of this brouhaha will not be expanded coverage to include creationism (that would also make me sad), but the reduction or excision of evolution from high school curricula. Evolution is one of the half dozen "great ideas" developed by science. It speaks to the profound issues of genealogy that fascinate all of us—the "roots" phenomenon writ large. Where did we come from? Where did life arise? How did it develop? How are organisms related? It forces us to think, ponder, and wonder. Shall we deprive millions of this knowledge and once again teach biology as a set of dull and unconnected facts, without the thread that weaves diverse material into a supple unity?

But most of all I am saddened by a trend I am just beginning to discern among my colleagues. I sense that some now wish to mute the healthy debate about theory that has brought new life to evolutionary biology. It provides grist for creationist mills, they say, even if only by distortion. Perhaps we should lie low and rally round the flag of strict Darwinism, at least for the moment—a kind of old-time religion on our part.

But we should borrow another metaphor and recognize that we too have to tread a straight and narrow path, surrounded by roads to perdition. For if we ever begin to suppress our search to understand nature, to quench our own intellectual excitement in a misguided effort to present a united front where it does not and should not exist, then we are truly lost.

QUESTIONS FOR DISCUSSION

1. Why, according to Gould, is "scientific creationism" a "meaningless and self-contradictory phrase"?
2. Gould says that he can "envision observations and experiments that would disprove any evolutionary theory" he knows. What might some such observations/experiments be?
3. Do you agree with Gould that arguments based on inference can be—and in the case of evolution are—just as "secure" as those based on "direct observation"?
4. Gould says that "all historical sciences rest upon inference." Should we infer that inference is not involved in non-historical sciences?
5. Why is Gould "most . . . saddened" by what he perceives as a trend among his scientific colleagues to present a "united front" against the creationists? In what sense would "we" be "truly lost"?

MAX WERTHEIMER

Einstein: The Thinking That Led to the Theory of Relativity*

The philosopher and psychologist Max Wertheimer was born in Prague and educated there and in Germany, receiving his Ph.D. from the University of Würzburg in 1904. After teaching for twelve years at the University of Berlin, he emigrated to the United States in 1934. From that year until his death in 1943, he was Professor of Psychology and Philosophy at the New School for Social Research in New York.

As co-founder of Gestalt theory, Wertheimer had a profound influence on twentieth-century thought. Indeed, his insights into the processes of perception changed our understanding of thinking itself. In a series of brilliantly conceived and scientifically exacting experiments, he was able to show, for example, that the parts of a given process (including the process of perception) cannot be understood independently of their place in the whole. For Wertheimer, this notion had special relevance for scientific investigation, with its tendency to analyze the parts of a system in isolation.

In the following essay, taken from Productive Thinking *(1945), Wertheimer dramatizes the thought processes by which Einstein struggled toward the theory that recast modern physics and reshaped our understanding of the universe. This piece goes well with Popper (this Part) and with Kneller and Bauer (Part I).*

Those were wonderful days, beginning in 1916, when for hours and hours I was fortunate enough to sit with Einstein, alone in his study, and hear from him the story of the dramatic developments which culminated in the theory of relativity. During those long discussions I questioned Einstein in great detail about the concrete events in his thought. He described them to me, not in generalities, but in a discussion of the genesis of each question.

Einstein's original papers give his results. They do not tell the story of his thinking. In the course of one of his books he did report some steps in the process. I have quoted him in the proper places in this [essay].

The drama developed in a number of acts.

Act I. The Beginning of the Problem

The problem started when Einstein was sixteen years old, a pupil in the Gymnasium (Aarau, Kantonschule). He was not an especially good student, unless he did productive work on his own account. This he did in physics and mathematics, and consequently he knew more about these subjects than his classmates. It was then that the great problem really started to trouble him. He was intensely concerned with it for seven years; from the moment, however, that he came to question the customary concept of time (see Act VII), it took him only five weeks to write his paper on relativity—although at this time he was doing a full day's work at the Patent Office.

The process started in a way that was not very clear, and is therefore difficult to describe—in a certain state of being puzzled. First came such questions as: What if one were to run after a ray of light? What if one were riding on the beam? If one were to run after a ray of light as it travels, would its velocity thereby be decreased? If one were to run fast enough, would it no longer move at all? . . . To young Einstein this seemed strange.

The same light ray, for another man, would have another velocity. What is "the velocity of light"? If I have it in relation to something, this value does not hold in relation to something else which is itself in motion. (Puzzling to think that under certain conditions light should go more quickly in one direction than another.) If this is correct, then consequences would also have to be drawn with reference to the earth, which is moving. There would then be a way of finding out by experiments with light whether one is in a moving system! Einstein's interest was captured by this; he tried to find methods by which it would be possible to establish or to measure the movement of the earth—and he learned only later that physicists had already made such experiments. His wish to design such experiments was always accompanied by some doubt that the thing was really so; in any case, he felt that he must try to decide.

He said to himself: "I know what the velocity of a light ray is in relation to a system. What the situation is if another system is taken into account seems to be clear, but the consequences are very puzzling."

Act II. Light Determines a State of Absolute Rest?

Would operations with light lead to conclusions different in this respect from conclusions from mechanical operations? From the point of view of mechanics there seems to be no absolute rest; from the point of view of light there does seem to be. What of the velocity of light? One must relate it to something. Here the trouble starts. Light determines a state of absolute rest? However, one does not know whether or not one is in a moving system. Young Einstein had reached some kind of conviction that one cannot notice whether or not one is in a moving system; it seemed to him deeply founded in nature that there is no "absolute movement." The central point here became the conflict between the view that light velocity seems to presuppose a state of "absolute rest" and the absence of this possibility in the other physical processes.

Back of all this there had to be something that was not yet grasped, not yet understood. Uneasiness about this characterized young Einstein's state of mind at this time.

When I asked him whether, during this period, he had already had some idea of the constancy of light velocity, independent of the movement of the reference system, Einstein answered decidedly: "No, it was just curiosity. That the velocity of light could differ depending upon the movement of the observer was somehow characterized by doubt. Later developments increased that doubt." Light did not seem to answer when one put such questions. Also light, just as mechanical processes, seemed to know nothing of a state of absolute movement or of absolute rest. This was interesting, exciting.

Light was to Einstein something very fundamental. At the time of his studies at the Gymnasium, the ether was no longer being thought of as something mechanical, but as "the mere carrier of electrical phenomena."

Act III. Work on the One Alternative

Serious work started. In the Maxwell equations of the electromagnetic field, the velocity of light plays an important role; and it is constant. If the Maxwell equations are valid with regard to one system, they are not valid in another. They would have to be changed. When one tries to do so in such a way that the velocity of light is not assumed to be constant, the matter becomes very complicated. For years Einstein tried to clarify the problem by studying and trying to change the Maxwell equations. He did not succeed in formulating these equations in such a way as to meet the difficulties satisfactorily. He tried hard to see clearly the relation between the velocity of light and the facts of movement in mechanics. But in whatever way he tried to unify the question of mechanical movement with the electromagnetic phenomena, he got into difficulties. One of his questions was: What would happen to the Maxwell equations and to their agreement with the facts if one were to assume that the velocity of light depends on the motion of the source of the light?

The conviction grew that in these respects the situation with regard to light could not be different from the situation with regard to mechanical processes (no absolute movement, no absolute rest). What took him so much time was this: he could not doubt that the velocity of light is constant and at the same time get a satisfactory theory of electromagnetic phenomena.

Act IV. Michelson's Result and Einstein

The famous Michelson experiment confronted physicists with a disconcerting result. If you are running away from a body that is rushing toward you, you will expect it to hit you somewhat later than if you are standing still. If you run toward it, it will hit you earlier. Michelson did just this in measurements of the velocity of light. He compared the time light takes to travel in two pipes if these pipes meet at right angles to each other, and if one lies in the direction of the movement of the earth, while the other is vertical to it. Since the first pipe, in its lengthwise direction, is moving with the movement of the earth, the light traveling in it ought to reach the receding end of this pipe later than the light in the other pipe reaches its end. . . .

No difference was found. The experiment was repeated, and the negative result was clearly confirmed.

The result of Michelson experiment in no way fitted the fundamental ideas of the physicists. In fact the result contradicted all their reasonable expectations.

For Einstein, Michelson's result was not a fact for itself. It had its place within his thoughts as they had thus far developed. Therefore, when Einstein read about these crucial experiments made by physicists, and the finest ones made by Michelson, their results were no surprise to him, although very important and decisive. They seemed to confirm rather than to undermine his ideas. But the matter was not yet entirely cleared up. Precisely how does this result come about? The problem was an obsession with Einstein although he saw no way to a positive solution.

Act V. The Lorentz Solution

Not only Einstein was troubled; many physicists were. Lorentz, the famous Dutch physicist, had developed a theory which formulated mathematically what had occurred in the Michelson experiment. In order to explain this fact it seemed necessary to him, as it had to Fitzgerald, to introduce an auxiliary hypothesis: he assumed that the entire apparatus used in the measurement underwent a small contraction in the direction of the earth's motion. According to this theory, the pipe in the direction of the movement of the earth was changed in length, while the other pipe suffered only a change in width and the length remained unaffected. The contraction had to be assumed to be just the amount needed to compensate for the effect of the earth's motion on the traveling of the light. This was an ingenious hypothesis.

There was now a fine, positive formula, determining the Michelson results mathematically, and an auxiliary hypothesis, the contraction. The difficulty was "removed." But for Einstein the situation was no less troublesome than before; he felt the auxiliary hypothesis to be a hypothesis *ad hoc*, which did not go to the heart of the matter.

Act VI. Re-examination of the Theoretical Situation

Einstein said to himself: "Except for that result, the whole situation in the Michelson experiment seems absolutely clear; all the factors involved and their interplay seem clear. But are they really clear? Do I really understand the structure of the whole situation, especially in relation to the crucial result? During this time he was often depressed, sometimes in despair, but driven by the strongest vectors.

In his passionate desire to understand or, better, to see whether the situation was really clear to him, he faced the essentials in the Michelson situation again and again, especially the central point: the measurement of the speed of light under conditions of movement of the whole set in the crucial direction.

This simply would not become clear. He felt a gap somewhere without being able to clarify it, or even to formulate it. He felt that the trouble went deeper than the contradiction between Michelson's actual and the expected result.

He felt that a certain region in the structure of the whole situation was in reality not as clear to him as it should be, although it had hitherto been accepted without question by everyone, including himself. His proceeding was somewhat as follows: There is a time measurement while the crucial movement is taking place. "Do I see clearly," he asked himself, "the relation, the inner connection between the two, between the measurement of time and that of movement? Is it clear to me how the measurement of time works in such a situation?" And for him this was not a problem with regard to the Michelson experiment only, but a problem in which more basic principles were at stake.

Act VII. Positive Steps toward Clarification

It occurred to Einstein that time measurement involves simultaneity. What of simultaneity in such a movement as this? To begin with, what of simultaneity of events in different places?

He said to himself: "If two events occur in one place, I understand clearly what simultaneity means. For example, I see these two balls hit the identical goal at the same time. But . . . am I really clear about what simultaneity means when it refers to events in two different places? What does it mean to say that this event occurred in my room at the same time as another event in some distant place? Surely I can use the concept of simultaneity for different places in the same way as for one and the same place—but can I? Is it as clear to me in the former as it is in the latter case? . . . It is not!"

For what now followed in Einstein's thinking we can fortunately report paragraphs from his own writing. He wrote them in the form of a discussion with the reader. What Einstein here says to the reader is similar to the way his thinking proceeded: "Lightning strikes in two distant places. I assert that both bolts struck simultaneously. If now I ask you, dear reader, whether this assertion makes sense, you will answer, 'Yes, certainly.' But if I urge you to explain to me more clearly the meaning of this assertion, you will find after some deliberation that answering this question is not as simple as it at first appears.

"After a time you will perhaps think of the following answer: 'The meaning of the assertion is in itself clear and needs no further clarification. It would need some figuring out, to be sure, if you were to put me to the task of deciding by observation whether in a concrete case the two effects were actually simultaneous or not.'"

I now insert an illustration which Einstein offered in a discussion. Suppose somebody uses the word "hunchback." If this concept is to have any clear meaning, there must be some way of finding out whether or not a man has a hunched back. If I could conceive of no possibility of reaching such a decision, the word would have no real meaning for me.

"Similarly," Einstein continued, "with the concept of simultaneity. The concept really exists for the physicist only when in a concrete case there is some possibility of deciding whether the concept is or is not applicable. Such a definition of simultaneity is required, therefore, as would provide a method for deciding. As long as this requirement is not fulfilled, I am deluding myself as physicist (to be sure, as non-physicist too!) if I believe that the assertion of simultaneity has a real meaning. (Until you have truly agreed to this, dear reader, do not read any further.)

"After some deliberation you may make the following proposal to prove whether the two shafts of lightning struck simultaneously. Put a set of two mirrors, at an angle of 90° to each other, at the exact halfway mark between the two light effects, station yourself in front of them, and observe whether or not the light effects strike the mirrors simultaneously."

Simultaneity in distant places here gets its meaning by being based on clear simultaneity in an identical place.

All these steps came not by way of isolated clarification of this special question, but as part of the attempt to understand the inner connection that was mentioned above, the problem of the measurement of speed during the crucial movement. In the mirror situation this means simply: What happens if, in the time during which the light rays approach my mirrors, I move with them, away from one source of light and toward the other? Obviously, if the two events appeared simultaneous to a man at rest they would not then appear so to me, who am moving with my mirrors. His statement and mine must differ. We see then that our statements about simultaneity *involve essentially reference to movement of the observer.* If simultaneity in distant places is to have real meaning, I must explicitly take into account the question of movement, and in comparing my judgments with those of another observer, I have to take into account the relative movement between him and me. When dealing with "simultaneity in different places" I must refer to the relative movement of the observer.

I repeat: suppose that I with my mirrors am traveling in a train going in a straight line at a constant velocity. Two shafts of lightning strike in the distance, one near the engine, the other near the rear end of the train, my double mirror being right in the middle between the two. As a passenger I use the train as my frame of reference, I relate these events to the train. Let us assume that just at the critical moment when the lightning strikes, a man is standing beside the tracks, likewise with double mirrors, and that his place at that moment coincides with mine. What would my observations be and what would his be?

"If we say that the bolts of lightning are simultaneous with regard to the tracks, *this now means:* the rays of light coming from two equidistant points meet simultaneously at the mirrors of the man on the track. But if the place of my moving mirrors coincides with his mirrors at the moment the lightning strikes, the rays will not meet exactly simultaneously in my mirrors because of my movement.

"Events which are simultaneous in relation to the track are not simultaneous in relation to the train, and vice versa. Each frame of reference, each system of coordinates therefore has *its special time;* a statement about a time has real meaning only when the frame of reference is stated, to which the assertion of time refers."

It has always seemed simple and clear that a statement about the "time difference" between two events is a "fact," independent of other factors, such as movement of the system. But, in actual fact, is not the thesis that "the time difference between two events is independent of the movement of the system" an arbitrary assumption? It did not hold, as we saw, for simultaneity in different places, and therefore it cannot hold even for the length of a second. To measure a time interval, we must use a clock or the equivalent of a clock, and look for certain coincidences at the beginning and at the end of the interval. Therefore the trouble with simultaneity is involved. We cannot dogmatically assume that the time which a certain event takes in relation to the train is the same as the time in relation to the track.

This applies also to the measurement of distances in space! If I try to measure exactly the length of a car by marking its end points on the roadbed, I must take care, when I have made my mark at one end, that the car does not move before I come to the other end! Unless I have explicitly given attention to this possibility, my measurements will be misleading.

I must therefore conclude that in every such measurement reference must be made to the movement of the system. For the observer within the moving system will get results which differ from those of an observer in another frame of reference. "Every system has its special time and space values. A time or space judgment has sense only if we know the system with reference to which the judgment was made." We must change the old view: the measurements of time intervals and of distances in space are not independent of the conditions of movement of the system in relation to the observer.

The old view had been a time-honored "truth." Einstein, seeing that it was questionable, came to the conclusion that space and time measurements depend on the movement of the system.

Act VIII. Invariants and Transformation

What followed was determined by two vectors which simultaneously tended toward the same question.

1. The system of reference may vary; it can be chosen arbitrarily. But in order to reach physical realities, I have to get rid of such arbitrariness. The basic laws must be independent of arbitrarily chosen coordinates. If one wants to get a description of physical events, the basic laws of physics must be invariant with regard to such changes.

Here it becomes clear that one might adequately call Einstein's theory of relativity just the opposite, an absolute theory.

2. Insight into the interdependence of time measurement and movement is certainly not enough in itself. What is now needed is a transformation formula that answers this question: "How does one find the place and time values of an event in relation to one moving system, if one knows the places and times as measured in another? Or better, how does one find the transformation from one system to another when they move in relation to each other?"

What would be the direct way? In order to proceed realistically, I would have to base the transformation on an assumption with regard to some physical realities which could be used as invariants.

The reader may think back to an old historical situation. Physicists in past ages tried to construct a *perpetuum mobile*. After many attempts which did not succeed, the question suddenly arose: how would physics look if nature were basically such as to make a *perpetuum mobile* impossible? This involved an enormous change, which recentered the whole field.

Similarly there arose in Einstein the following question, which was inspired by his early ideas mentioned in Acts II and III. How would physics look if, by nature, measurements of the velocity of light would under all conditions have to lead to the identical value? Here is the needed invariant! (Thesis of the basic constancy of the velocity of light.)

In terms of the desired transformation, this means: "Can a relation between the place and time of events in systems which move linearly to each other be so conceived that the velocity of light becomes a constant?"

Eventually Einstein reached the answer: "Yes!" The answer consisted of concrete and definite transformation formulas for distances in time and space, formulas that differed characteristically from the old Galilean transformation formulas.

3. In the discussions I had with Einstein in 1916 I put this question to him: "How did you come to choose just the velocity of light as a constant? Was this not arbitrary?"

Of course it was clear that one important consideration was the empirical experiments which showed no variation in the velocity of light. "But did you choose this arbitrarily," I asked, "simply to fit in with these experiments and with the Lorentz transformation?" Einstein's first reply was that we are entirely free in choosing axioms. "There is no such difference as you just implied," he said, "between reasonable and arbitrary axioms. The only virtue of axioms is to furnish fundamental propositions from which one can derive conclusions that fit the facts." This is a formulation that plays a great role in present theoretical discussions, and about which most theorists seem to be in agreement. But then Einstein himself smilingly proceeded to give me a very nice example of an unreasonable axiom: "One could of course choose, say, the velocity of sound instead of light. It would be reasonable, however, to select not the velocity of just 'any' process, but of an 'outstanding' process. . . ." Questions like the following had occurred to Einstein: Is the speed of light perhaps the fastest possible? Is it perhaps impossible to accelerate any movement beyond the speed of light? As velocity increases, progressively greater forces are required to increase it still further. Perhaps the force required to increase a velocity beyond the velocity of light is infinite?

It was marvelous to hear in Einstein's descriptions how these bold questions and expectations had taken shape in him. It was new, unthought of before, that the velocity of light might be the greatest possible velocity, that an attempt to go beyond that limit would require forces infinitely great.

If these assumptions brought clarity into the system, and if they were proved by experiment, then it would make good sense to take the velocity of light as the basic constant. (Cf. the absolute zero of temperature which is reached when the molecular movements in an ideal gas approach zero.)

4. The derivations which Einstein reached from his transformation formulas showed mathematical coincidence with the Lorentz transformation. The contraction hypothesis had therefore been in the right direction, only now it was no longer an arbitrary auxiliary hypothesis, but the outcome of improved insight, a logically necessary derivation from the improved view of fundamental physical entities. The contraction was not an absolute event, but a result of the relativity of measurements. It was not determined by a "movement in itself which possesses no real sense for us, but only by a movement with reference to the chosen observation system."

Act IX. On Movement, on Space, a Thought Experiment

The last statement throws new light on the changes in thinking which were already involved in the earlier steps. "By the motion of a body we always mean its change of position in relation to a second body," to a framework, or a system. If there is one body alone, it makes no sense to ask or to try to state whether it is moving or not. If there are two, we can state only whether they are approaching or moving away from each other, but, so long as there are only two, it makes no sense to ask, or to try to state, whether one is turning around the other; the essential in movement is change of position in relation to another object, a framework, or a system.

But is there not one outstanding system in regard to which there is *absolute* movement of a body, "the" space (Newtonian space, the space of the ether), the box in which all movement takes place?

Here I may mention something that happened not just at this point in the development of the process, but may illustrate what was really going on. It transcends the problems of the special theory of relativity: Is there no proof of the reality of such an outstanding system? A famous experiment of Newton's had been used as proof: When a sphere of oil rotates it becomes flattened. This is a real, physical, observable fact, apparently caused by an "absolute" movement.

But is this really a demonstration of such an absolute movement? It seems so certainly; but is it actually, if we think it through? In reality we have not a body moving alone in absolute space, but a body that moves within our fixed-star firmament. Is the flattening of that sphere perhaps an outcome of the movement of the sphere relative to the surrounding stars? What would happen if we took a very huge iron wheel, with a small hole at the center, if we suspended in this hole a little sphere of oil, and then rotated the wheel? Perhaps the little sphere would again become flattened. Then the flattening would have nothing to do with the rotation in an absolute space box; rather it would be determined by the systems moving in relation to each other, the big wheel or the firmament on the one hand and the little sphere of oil on the other.

Of course rotation already transcends the region of the so-called special relativity of Einstein. It became basic in the problem of the general theory of relativity.

Act X. Questions for Observation and Experiment

Einstein was at heart a physicist. Thus all these developments aimed at real, concrete, experimental problems. As soon as he reached clarification he concentrated on the point: "Is it possible to find crucial physical questions to be answered in experiments that will decide whether these new theses are 'true'; whether they fit facts better, give better predictions of physical events than the old theses?"

He found a number of such crucial experiments, some of which physicists could and later did carry out.

* * *

Let us recapitulate briefly.

First there was what we may call the foreperiod. Einstein was puzzled by the question, first, of the velocity of light when the observer is in motion. He considered, secondly, the consequences as to the question of "absolute rest." Thirdly, he then tried to make one alternative workable (is the velocity of light in Maxwell's equations a variable?), and obtained a negative result. There was, fourth, the Michelson experiment which confirmed the other alternative—and, fifth, the Lorentz-Fitzgerald hypothesis, which did not seem to go to the root of the trouble.

So far everything, including the meaning and structural role of time, space, measurement, light, etc., was understood in terms of traditional physics—structure I.

In this troubled situation the question arose: Is this structure itself, in which the Michelson result seems contradictory, really clear to me? This was the revolutionary moment. Einstein felt that the contradiction should be viewed without prejudice, that the time-honored structure should be requestioned. Was this structure I adequate? Was it clear just with regard to the critical point—the question of light in relation to the question of movement? Was it clear in the situation of the Michelson experiment? All these questions were asked in a passionate effort to understand. And then the procedure became more specific in one step after another.

How was the velocity of light to be measured in a moving system?

How was time to be measured under these circumstances?

What does simultaneity mean in such a system?

But, then, what does simultaneity mean if the term is referred to different places?

The meaning of simultaneity was clear if two events occur in the same place. But Einstein was suddenly struck by the fact that it was not equally clear for events in distant places. Here was a gap in any real understanding. He saw: It is blind simply to apply the customary meaning of simultaneity to these other cases. If simultaneity is to have a real meaning, we must raise the question of its factual recognition so that in concrete cases we can tell whether or not the term applies. (Clearly, this was a fundamental logical problem.)

The meaning of simultaneity in general had to be based on the clear simultaneity in the case of spatial coincidence. But this required that in every case of different location of two events the relative movement be taken into account. Thus the meaning, the structural role of simultaneity in its relation to movement, underwent a radical change.

Immediately, corresponding requirements follow for the measurement of time in general, for the meaning, say, of a second, and for the measurement of space, since they must now depend upon relative movement. As a result, the concepts of time-flow, of space, and of the measurement both of time and space change their meaning radically.

At this point the introduction of the observer and his system of coordinates seemed to introduce a fundamentally arbitrary or subjective factor. "But the reality," Einstein felt, "cannot be so arbitrary and subjective." In his desire to get rid of this arbitrary element and, at the same time, to get a concrete transformation formula between various systems, he realized that a basic invariant was needed, some factor that remains unaffected by the transition from one system to another. Obviously, both demands went in the same direction.

This led to the decisive step—the introduction of the velocity of light as the invariant. How would physics look if recentered with this as a starting point? Bold consequences followed one after another, and a new structure of physics was the consequence.

When Einstein reached the concrete transformation formula on the basis of this invariant, the Lorentz transformation appeared as a derivation—but now it was understood in a deeper, entirely new way, as a necessary formulation within the new structure of physics. The Michelson result, too, was now seen in an entirely new light, as a necessary result when the interplay of all relative measurements within the moving system was taken into account. Not the result was troublesome—he had felt that from the very beginning—but the behavior of the various items in the situation before finding the solution. With the deeper understanding of these items the result was required.

The picture was now improved. Einstein could proceed to the question of experimental verification.

In the briefest formulation: In a passionate desire for clearness, Einstein squarely faced the relation between the velocity of light and the movement of a system, and confronted the theoretical structure of classical physics and the Michelson result.

A part-region in this field became crucial and was subjected to a radical examination.

Under this scrutiny a great gap was discovered (in the classical treatment of time).

The necessary steps for dealing with this difficulty were realized.

As a result, the meaning of all the items involved underwent a change.

When a last arbitrariness in the situation had been eliminated, a new structure of physics crystallized.

Plans were made to subject the new system to experimental test.

Radical structural changes were involved in the process, changes with regard to separateness and inner relatedness, grouping, centering, etc.; thereby deepening, changing the meaning of the items involved, their structural role, place, and function in the transition from structure I to structure II. It may be advisable to explain once more in what sense Einstein's achievement meant a change of structure.

1. In the Michelson situation—as in classical physics generally—time had been regarded as an independent variable and, therefore, as an independent tool in the business of measurement, entirely separate from, in no way functionally interdependent with the movements that were involved in that observational situation. Accordingly, the nature of time had been of no interest with regard to the apparently paradoxical result.

In Einstein's thought there arose an intimate relationship between time-values and the physical events themselves. Thus the role of time within the structure of physics was fundamentally altered.

This radical change was first clearly envisaged in the consideration of simultaneity. In a way, simultaneity split in two: the clear simultaneity of events in a given place and, related to it, but related by means of specific physical events, the simultaneity of events in different places, particularly under conditions of movement of the system.

2. As a consequence, space-values also changed their meaning and their role in the structure of physics. In the traditional view they, too, had been entirely separated from, independent of time and of physical events. Now an intimate relation was established. Space was no longer an empty and wholly indifferent container of physical facts. Space geometry became integrated with the dimension of time in a four-dimensional system, which in turn formed a new unitary structure with actual physical occurrences.

3. The velocity of light had so far been one velocity among many. Although the highest velocity known to the physicist, it had played the same role as other velocities. It had been fundamentally unrelated to the way in which time and space are measured. Now it was considered as closely bound up with time- and space-values, and as a fundamental fact in physics as a whole. Its role changed from that of a particular fact among many to that of a central issue in the system.

Many more items could be mentioned which changed their meaning in the process, such as mass and energy, which now proved to be closely related. But it will not be necessary to discuss further particulars.

In appraising these transformations we must not forget that they took place in view of a gigantic given system. Every step had to be taken against a very strong gestalt—the traditional structure of physics, which fitted an enormous number of facts, apparently so flawless, so clear that any local change was bound to meet with the resistance of the whole strong and well-articulated structure. This was probably the reason why it took so long a time—seven years—until the crucial advance was made.

One could imagine that some of the necessary changes occurred to Einstein by chance, in a procedure of trial and error. Scrutiny of Einstein's thought always showed that when a step was taken this happened because it was required. Quite generally, if one knows how Einstein thought, one knows that any blind and fortuitous procedure was foreign to his mind.

The only point at which there could have been some doubt in this respect was the introduction of the constancy of light velocity in Einstein's general transformation formulas. In a thinker of lesser stature this could have happened through mere tentative generalization of the Lorentz formula. But actually the essential step was not reached in this fashion; there was no mathematical guesswork in it.

In late years Einstein often told me about the problems on which he was working at the time. There was never a blind step. When he dropped any direction, it was only because he realized that it would introduce ununderstandable, arbitrary factors. Sometimes it happened that Einstein was faced with the difficulty that the mathematical tools were not far enough developed to allow a real clarification; nonetheless he would not lose sight of his problem and would often succeed in finding a way eventually, in which the seemingly insuperable difficulties could be surmounted.

QUESTIONS FOR DISCUSSION

1. Describe Michelson's experiment with light. What was the expected result? What was the actual result?
2. How did Lorentz explain Michelson's "surprising" result? Why was Einstein dissatisfied with Lorentz's hypothesis?
3. Explain Einstein's revised understanding of "simultaneity." How did this new understanding affect his/our understanding of "space" and "time"?
4. Do you find it surprising that Einstein's Theory of Relativity is based on an "absolute"? Why was an absolute ("invariant") necessary?
5. Wertheimer says that "Einstein was at heart a physicist"; thus his work "aimed at real, concrete, experimental problems." What do you take to be the force of this statement?

Part III

Contexts of Discovery

Introduction

Discovery is essential to the progress of science, and yet it is a remarkable fact that when scientists embark on a research problem (or apply for funds to carry it out) they never do so with the assurance that an important discovery will follow. Rather, their research is typically justified as an "investigation" and presented in the form of a "problem." Discoveries, if they come, frequently come out of the blue and may have only a tangential (and sometimes no) relationship to the immediate problem under investigation. Wilhelm Röntgen's discovery of X-rays, for example, was an unexpected result of his investigation into the behavior of electric discharges in high vacua. And Ernest Rutherford's discovery of the atomic nucleus followed from the surprising result of Hans Geiger and Ernest Marsden's experiment designed to tell them something about alpha particles. Both discoveries, in other words, were unforeseen consequences of apparently unrelated investigations. The history of science is rife with such "accidental" discoveries—among the most famous being Alexander Fleming's discovery of penicillin, Louis Pasteur's discovery of cholera vaccine, and Jocelyn Bell Burnell's discovery of pulsars.

This unpredictable and apparently accidental quality of many important discoveries tells us something about both nature and science. Perhaps most importantly, it reminds us that nature's order is subject neither to the dictates of the human intellect nor to restrictions imposed by the limitation of our sensory perceptions. We can bring the most rigorous logic and creative imagination—as well as our most acute powers of observation (enhanced by sophisticated technology)—to bear, but they cannot guarantee that we will arrive at nature's truth. We start down one carefully defined path of inquiry only to find that it has led in an entirely unexpected direction. Because nature operates according to its own laws, independent of the exigencies of the human mind, it has a way of thwarting our expectations, of throwing up surprises.

What does this unpredictable quality of discovery tell us about science and its ways of knowing? One thing it suggests is that traditional notions of how we arrive at scientific knowledge (strict logic, careful experimentation, systematic verification, etc.) are only part of the story. Certainly, science requires its "proofs." But proofs have to do with "answers," with confirming what we (think we) know. Science most fundamentally, however, seeks to *expand* our knowledge, to explore the unknown—to discover. And it does so by searching out and pursuing problems, that is, anomalies, conditions, results, etc. not explainable by the current state of scientific knowledge. Because such problems involve the unexpected and/or unexplained, we cannot be surprised that the path to their resolution is not always clearly marked. As the distinguished biologist Peter Medawar has reminded us, there is "no procedure of discovery that can be logically scripted." Rather, the scientist uses a variety of "exploratory stratagems" that have as much to do with personal style as with logic. Acute powers of observation, a delight in surprise, a willingness to grapple with the unexpected, and a "feel for things" are but some of these stratagems.

Part III explores some of the contexts and conditions in which discovery takes place. In the first essay, Robert S. Root-Bernstein argues that discovery often results not from the logical process of verification but rather from problems or anomalies exposed during the course of verification. For him, our understanding of the way science is actually done must be broadened to account for the "messy process of discovery." In the essay that follows, Mahlon Hoagland discusses discovery from a different perspective. He argues that because science is a "dynamic accumulating body of knowledge," and because it is an essentially collaborative enterprise, we can assume that if any given discovery had not been made by one scientist, it would sooner or later have been made by another (or others). A discovery, he believes, is "uniquely the discoverer's only in terms of priority and in the way it was made." These issues of uniqueness and collaboration in scientific discovery are implicit in Anne Sayre's examination of Rosalind Franklin's role in the discovery of the double-helix structure of DNA. Sayre discusses the ethical issues surrounding this discovery and shows that they are rooted in the inherent contradiction between the essentially collaborative and the inevitably competitive nature of science. These contradictory characteristics loom large in Luis W. Alvarez's account of his near misses in the discovery of nuclear fission and other breakthroughs in atomic physics. In the next essay, Evelyn Fox Keller, in her discussion of Nobel Laureate Barbara McClintock, returns to issues raised earlier by Root-Bernstein. As Keller shows, McClintock's greatness as a scientist lay in her ability to achieve a remarkable intimacy with her material ("a feeling for the organism"). Intimacy implies a collapsing of the distance between the observer and the observed (between subject and object). For McClintock, as for a number of the scientists discussed by Root-Bernstein, scientific objectivity (and, more broadly, scientific method) as traditionally understood limits our apprehension of nature's mysteries. The importance of this deep familiarity with one's material is humorously underscored by Samuel Scudder in the penultimate essay. In the final selection of Part III, the Poinars convey something of the excitement—and deep humility—that accompanied their discovery of soft tissue (intact cells) in ancient amber. For the Poinars, as for so many of the scientists represented/discussed in this collection, science involves far more than the purely rational self. Nature's mysteries demand our reverence as well as our reason.

ROBERT S. ROOT-BERNSTEIN

Setting the Stage for Discovery*

A professor of natural science and physiology at Michigan State University, Robert Root-Bernstein studies the causes of autoimmune diseases and the interactions between drugs and neurotransmitters. In 1989 he published Discovering, *a book on the strategies of scientific discovery.*

---

*Originally appeared in *The Sciences*, May/June 1988. Reprinted by permission of the publisher.

His Rethinking AIDS: The Tragic Cost of Premature Consensus *appeared in 1993. Root-Bernstein was one of the first MacArthur Prize Fellows and was for several years a contributing editor of* The Sciences, *published by the New York Academy of Sciences.*

In the following essay, Root-Bernstein broadens the traditional definition of scientific method to include a range of attitudes and activities that can set the stage for discovery. He stresses the importance of intuition and even of deep personal engagement as a means of achieving scientific insight. On this, see also Keller and the Poinars (this Part).

Anyone familiar with the history or philosophy of science has heard some version of the story in which a researcher is going patiently about his daily grind—growing cell cultures or mixing chemicals or peering into a microscope—when, quite by accident, he makes some earthshaking discovery. Variations on the tale are myriad, but the moral is always the same: Great breakthroughs can be neither planned nor predicted; you just have to get lucky.

Consider, for example, the legend of how Louis Pasteur developed the cholera vaccine. According to the standard account, the French chemist might never have realized that weakened microorganisms can activate the immune system without causing serious illness had he not gone away on vacation during the summer of 1879. Pasteur had been experimenting with chicken cholera, the story goes, and happened to leave his germ cultures sitting out when he left Paris for more than two months. Upon his return, he found that the cultures, though still active, had become avirulent; they no longer could sicken a chicken. So he developed a new set of cultures from a natural outbreak of the disease and resumed his work. Yet he found, to his surprise, that the hens he had exposed to the weakened germ culture still failed to develop cholera. Only then did it dawn on Pasteur that he had inadvertently immunized them.

Equally fortuitous, according to conventional wisdom, was the German pathologist Oskar Minkowski's 1889 discovery that diabetes stems from a disorder of the pancreas. Minkowski had removed that organ from a dog to determine its role in the digestion of fat. After the operation, the dog happened to urinate on the laboratory floor. The urine drew flies, and the flies drew the attention of a sharp-eyed lab assistant. Puzzled, since flies are not normally attracted to urine, the assistant questioned Minkowski, who analyzed the urine and found it to be loaded with sugar. The obvious conclusion was that the pancreas was somehow involved in metabolizing that substance. (We now know that the pancreas contains islets of Langerhans, which secrete insulin, a hormone responsible for sugar metabolism.) The depancreatized dog turned out to be a perfect experimental model for diabetes. Yet, as legend has it, Minkowski would never have recognized this had the dog not relieved itself in the company of a swarm of flies and an alert lab assistant.

Still another breakthrough usually described as a fluke is the discovery of lysozyme—a bacteria-killing enzyme in tears, saliva, mucus, and other bodily fluids and tissues—by the British bacteriologist Alexander Fleming, in 1921. When Fleming began the work that led to this discovery, during the First World War, it was well known that the body had three lines of defense against infection—the skin (a physical barrier); macrophages (a type of white blood cell that ingests foreign material); and antibodies (proteins that neutralize toxins by adhering to them). But no one had even suggested there might be a fourth. Thus, we are told, Fleming was not looking for lysozyme; the discovery resulted from a series of chance occurrences. First, some contaminant from the air fell into a culture dish in Fleming's laboratory, where it spawned a bacterial colony. Then, when Fleming leaned over his microscope to take a close look at this germ population, his nose dripped into it (he suffered from frequent winter colds). To his surprise, the drippings dissolved colonies of bacteria in the petri dish. He developed other cultures from the first one, subjected them to the same treatment, and obtained the same result. Further experiments confirmed that the mucus contained an antibacterial agent and showed that the agent was a proteinaceous substance that did not reproduce itself. He concluded it was an enzyme manufactured by the body.

Stories such as these (there are countless others, ranging from Wilhelm C. Röntgen's discovery of the X-ray to Jocelyn Bell Burnell's discovery of pulsars to Charles R. Richet's discovery of anaphylaxis—an extreme allergic reaction that can cause death) have led philosophers of science to draw a bold distinction between the process of *discovery* and that of *proof,* and to insist that logic and reason apply only to the latter. According to most standard texts—W. I. B. Beveridge's *Art of Scientific Investigation,* R. B. Braithwaite's *Scientific Explanation,* Carl Hempel's *Philosophy of Natural Science,* David Hull's *Philosophy of Biological Science,* Karl Popper's *Logic of Scientific Discovery*—discovery is a product not of particular methods of logical inquiry but of being in the right place at the right time. It could happen to anyone at any time. In contrast, the process of testing a hypothesis is said to be a more logical operation—one that only a rational inquirer, trained in the methods of science, can successfully perform. Unlike discovery, scientific validation is thought to consist of two distinct mental activities: induction (deriving general rules from particular instances) and deduction (making specific predictions based on general rules). The objective of science, according to this philosophy, is simply to validate or invalidate inexplicable insights.

By limiting themselves to explaining scientific validation, philosophers save themselves the trouble of trying to account for the rich, messy business of discovery. But this approach has drawbacks: It suggests, paradoxically, that illogical processes have led to the most logical constructs known to mankind—mathematics and science. And it fails to explain where problems come from, what scientists do from day to day, and how they actually think. Real scientists do not spend their lives cataloguing the facts that follow from established principles, or noting the principles that are implicit in particular facts. As the French mathematician Henri Poincaré argued in *Science and Method,* such exercises would be largely pointless and sterile and endlessly boring. The passion of any real scientist is to expand our knowledge of the world, not merely to confirm it. That means searching out instances in which the codified rules of science fail to account for our experience: looking for paradoxes, contradictions, anomalies—in short, for problems. It is only after a problem has been identified that induction or deduction can serve a purpose, and only in relation to such a problem that an observation becomes a discovery.

So something is clearly amiss. The notion of accidental discovery assumes that anyone else seeing what Pasteur, Minkowski, Fleming, Röntgen, Burnell, or Richet saw would have come to the same conclusions. Yet, in each case, someone else *was* there and did *not* make the discovery: Pasteur's collaborator Émile Roux, Minkowski's unidentified lab assistant, and Fleming's colleague V. D. Allison. Richet reports that the experiment that caused him to invent the concept of anaphylaxis was so bizarre, his collaborators refused even to countenance the results. And several people, including the English chemist William Crookes, observed the same phenomena that Röntgen did—fogged photographic plates and fluorescing barium platinocyanide screens—but did not appreciate the fact that these effects were created by previously unknown rays—X-rays—emitted by nearby cathode-ray tubes. Clearly, it is not sufficient simply to be in the right place at the right time. How a scientist interprets what he sees depends on what he expects. Discoveries do not just walk up and present themselves from time to time, disguised as chickens, dogs, or nasal drips.

Why not admit that discoveries derive from the ways in which particular scientists logically go about their work? Then, given that different scientists practice different styles of research, and that not all of them make discoveries, it should be possible to identify the styles that most often pay off. Surely, any mental activity that contributes directly to scientific discoveries should be recognized as scientific method. If such activities are not acknowledged by the prevailing view of how scientists use logic and reason, that does not mean the activities are illogical. It means that the prevailing view is too narrow to account for how scientists really think. The task, then, is to redefine the scientific method in a way that accounts for the process of discovery.

Were Pasteur's and Minkowski's and Fleming's breakthroughs really just accidents? A recent analysis of Pasteur's notebooks by the historian of science Antonio Cadeddu suggests that the discovery of the cholera vaccine was anything but. It was well known during the late nineteenth century that people who survive certain infectious diseases tend not to come down with them again. Pasteur had noted as much, and his experiments with chicken cholera were clearly designed to explore that phenomenon. He seems to have been consciously pursuing a problem: how to produce a microbe strong enough to cause some degree of illness (and thus to protect against future infection) yet not strong enough to kill. So he was not aimlessly inoculating chickens when he discovered the cholera vaccine; he was trying, to use his own term, to "enfeeble" the infectious agent.

Moreover, the breakthrough did not come about from his leaving flasks of germs unattended while he went on vacation. In fact, he left them in the care of Émile Roux. Pasteur did, upon his return, inoculate chickens with material from the flasks, and the birds did fail to become ill. But when the same chickens were later injected with a more virulent strain, they died. No discovery here. Indeed, the notebooks reveal that Pasteur did not even initiate his first successful enfeeblement experiment until a few months later, in October of 1879. He and Roux had tried to enfeeble the germs by passing them from one animal to another, by growing them in different media, by heating them, by exposing them to air—anything that conceivably might weaken them—and only after many such attempts did one of the experiments succeed.

That winter, Pasteur managed, by placing germ cultures in acidic mediums, to enfeeble them in varying degrees. For some time, the strains that failed to kill chickens were also too weak to immunize them. But by March of 1880, Pasteur had developed two cultures with the properties of vaccines. The trick, according to his notebooks, was to use a mildly acidic medium, not a strong one, and to leave the germ culture sitting in it for a long time. Thus, he produced an attenuated organism capable of inducing an immune response in chickens. The discovery, therefore, was not an accident at all; Pasteur had posed a question—Is it possible to immunize an animal with a weakened infectious agent?—and then systematically searched for the answer.

Minkowski, too, was less reliant on dumb luck than is widely presumed. He probably *was* surprised to find sugar in the urine of the depancreatized dog; he had, after all, set out to investigate the role of the pancreas in the metabolism of fat. But Minkowski's own account of the discovery, published long after the popular version took hold, suggests it was neither a swarm of flies nor an alert lab assistant that brought the undigested sugar to his attention. It was, rather, his own carefully honed skills of observation and diagnosis, which he applied to an unexpected change in the dog's behavior.

Minkowski recounts that the dog, though fully housebroken before the operation, became an inveterate floor wetter afterward. In medical terms, it developed polyuria—unusually frequent urination. Polyuria is a classic symptom of diabetes, and Minkowski had learned in medical school that if a patient developed that symptom, the way to find out whether he had diabetes rather than, say, a bladder infection was to test the urine for sugar. Once Minkowski had asked the right question—namely, Why does a depancreatized dog suddenly develop polyuria?—standard medical procedures provided a ready answer: the urine was found to contain sugar and, as expected, the dog eventually developed all the symptoms of diabetes. So it followed that diabetes stems from a pancreatic disorder.

Minkowski's discovery was clearly a surprise (he had not even set out to study diabetes), but that is not to say it was a random occurrence. The dog's indoor accidents did not just happen; they were an inevitable result of the pancreas experiment. Nor was it by fluke that Minkowski found the dog's problem significant. His response was a consequence of his expectations. Had he not known the dog, he might have assumed that it always urinated on the floor. And had he not been familiar with the symptoms of diabetes, he might never have suspected that he had induced it in the dog. In short, Minkowski's discovery consisted not of what he saw but of how he saw it.

What about Fleming's discovery of lysozyme? According to the accepted accounts, there was no logic whatever to this breakthrough; it grew out of at least three totally unpredictable occurrences: first, Fleming got a severe cold; second, at about the same time, his petri dish was mysteriously contaminated by one of the few bacteria sensitive to lysozyme; third, he happened to contaminate the same dish with a drip from his nose and still did not discard it. In fact, one need only consult Fleming's notebooks to see that he quite literally cultivated the circumstances surrounding his discovery. The initial contaminant turns out to have been a bacterium harvested from his own nose—and his fateful drip into this culture, part of a deliberate experiment.

The purpose of the experiment was to determine whether colds might be caused by bacteriophages—viruses that cause illness by destroying resident bacteria in a host's body. The bacteriologist Frederick W. Twort had discovered bacteriophages in 1915, and a few years later, another bacteriologist, Félix d'Hérelle, isolated them in locusts with diarrhea and in humans with dysentery. The cold-prone Fleming was personally interested in learning the cause of the common cold, and a simple, flippant play on words may have led him to suspect bacteriophages. Might not "runny noses" and "runny bottoms" be the work of related agents? The way to find out was to extract bacteria from normal nasal mucus and then determine whether a cold sufferer's mucus contains agents capable of destroying it.

Fleming reported that it took him four days to cultivate a suitable bacterial colony. (The chance-drip version of the story is further belied by his assistant W. Howard Hughes's recollection, in *Alexander Fleming and Penicillin,* that Fleming had attached a leather guard to his microscope to prevent such accidents from occurring.) A few drops of his cold-infected mucus ate holes into this lawn of bacteria—just as bacteriophages do. Fleming spent the next few weeks conducting additional tests to make sure. But things started going wrong.

The easiest way to find out whether a solution contains bacteriophages is to dilute it repeatedly. Because bacteriophages are self-replicating, a solution containing them will return to its original potency within a few hours. Thus, when Fleming's solutions did not regain their strength, he began to suspect that he was dealing with an enzyme. (Because enzymes are body products, not organisms, they are not self-replicating; the more an enzyme preparation is diluted, the less activity there is.) Furthermore, Fleming found that the agent he had isolated could be inactivated by heat, as other enzymes can, and chemical tests demonstrated that it had the proteinaceous composition of an enzyme. The original hypothesis was foiled: the antibacterial agent obviously was not an invading bacteriophage. Instead, Fleming had discovered a new enzyme. He soon published his discovery, adding lysozyme to the pantheon of recognized bodily defenses. But he never publicly explained how he had happened upon lysozyme, and, hence, the story of the contaminated cell culture and the accidental drip was invented to make up for historical ignorance.

As Minkowski and Pasteur did, Fleming succeeded only after failing, but he did not succeed by chance. Had he not conceived of a possible link between intestinal disease and the common cold, he would not have been looking for bacteriophages in his nasal mucus. And had he not expected his would-be bacteriophages to reproduce in solution, he would not have performed the tests that led to his recognition of a new enzyme.

Virtually every so-called chance discovery that has been reexamined in the light of additional historical evidence has had to be revised in the manner of the Pasteur, Minkowski, and Fleming stories. Again and again, the record reveals that the discovery is not a fluke but the inevitable, if unforeseen, consequence of a rational and carefully planned line of inquiry initiated by a scientist. It follows that, contrary to philosophical orthodoxy, the tests of an incorrect hypothesis often result in surprises that lead to discovery, and that discoverers are not just beneficiaries of fate. They seem to have ways of courting the unexpected, which improve their chances of making novel observations. So there must be a logic, or at least a set of strategies, in discovery. The question is, Why are discoveries made by certain scientists rather than others? Can their strategies be learned?

I think they can. But such strategies are not so easily codified as are the rules of scientific proof, for they pertain to everything from recognizing interesting problems to appreciating unexpected results. How a scientist handles these matters is a function of his entire personality—the sum of the interests, skills, experiences, and desires that define him as a human being. Still, it should be possible to identify some of the habits of thought that are particularly advantageous.

It is striking how many great scientists have incorporated play into their lives and work, how many have consciously avoided being overly cautious or orderly, or narrowly dramatic. Fleming, for one, was famous for his love of games. He was raised in a family that played everything from poker and bridge to table tennis and quiz games. As an adult, he played croquet, bowls, and snooker at his club, and pitched pennies at his office whenever he lacked patients. He took up golf, too, but rarely played a straight game; he would putt holding the club as a snooker cue, or revise the rules to make the game more interesting. Life was essentially a game to him, and so was research. "I play with microbes," he once said, adding, "It is very pleasant to break the rules."

In the laboratory, one of Fleming's favorite pastimes was to fashion art from germs. He would start with an assortment of microorganisms and, knowing which color each one would produce as it multiplied, paint them onto a petri dish. After incubating the dish for a day, he would unveil a picture of his house or a ballerina or a mother nursing a baby. Fleming was no great artist, but his hobby fostered a rare intimacy with the bacterial world. To paint his pictures, he had to know not only which germ would produce which color but also how rapidly each would proliferate at a given temperature. To maintain a diverse palette, he also had to be constantly on the lookout for bacteria that might suit his purposes. To this end, he made a point of creating environments in which unusual germs might crop up. V. D. Allison recalls in a lecture to the Ulster Medical Society, just how conscientiously Fleming practiced this method:

> At the end of each day's work I cleaned my bench, put it in order for the next day and discarded tubes and culture plates for which I had no further use. He, for his part, kept his cultures. . . for two or three weeks until his bench was overcrowded with forty or fifty cultures. He would then discard them, first of all looking at them individually to see whether anything interesting or unusual had developed.

Fleming was not alone in his tendency to mix things up a bit to see what would happen. Konrad Lorenz, the great animal behaviorist, was equally scrupulous about cultivating fruitful confusion. Lorenz lived among his research subjects: dozens of species of mammals, birds, reptiles, and fishes. He did not quantify, control, or consciously experiment. He got to know each creature individually, then threw them together, watching for the unexpected, the unusual, or the bizarre in the chaos that followed. For example, his interest in one of ethology's most important concepts, that of intention movements (motions with meaning, such as the head bobbing in birds that serves as an alarm signal before flight), derived from an inadvertent experiment. He had trained a free-flying raven to eat raw meat from his hand and had been feeding the bird on and off for several hours one day. He would reach into his pants pocket and take out a piece of meat, and the raven would swoop down to grab it in its bill. By and by, Lorenz went to relieve himself near a hedge. When the raven saw him put his hand into his pants and pull out another morsel of meat, it swooped down, hungrily grasping the new mouthful in its bill. Lorenz howled in pain. But the event left a deep impression on him—about how faithfully animals respond to intention movements, that is.

One mental quality that facilitates discovery, then, is a willingness to goof around, to play games, and to cultivate a degree of chaos aimed at revealing the unusual or the unexpected. Looking back on the scientists who missed discoveries—Allison in Fleming's lab; Richet's collaborators on the anaphylaxis experiments; Crookes, Röntgen's colleague—we see that, in each case, they refused to credit a phenomenon with significance because it was not what they were looking for. "It's just a contamination." "You must have injected the wrong solution." "Send the photographic plates back to the manufacturers and tell them they'd better deliver good ones tomorrow or we'll cancel our order."

A classic example of such a reaction was reported by Jocelyn Bell Burnell in an interview concerning her discovery of pulsars. She had been pointing her radio telescope toward a region of the heavens at a time when she expected to pick up only a weak signal, when the pen on the recording device started jiggling. Repeating the observation at weekly intervals yielded the same result, and test after rest revealed nothing wrong with the equipment. Eventually, Burnell realized she had detected the presence of stellar sources of pulsating radio waves, or pulsars, which astronomers had hypothesized but never found. Sometime later, she heard that a colleague had observed the same phenomenon, given his equipment table a good kick, and written off the result as a mechanical aberration. We may presume he later kicked himself.

Not every anomaly or unexpected result leads to discovery, of course. As Sherlock Holmes once said, "It is of the highest importance in the art of detection to be able to recognize out of a number of facts which are incidental and which vital." However, Charles Richet, in *The Natural History of a Savant,* and the physicist George P. Thomson, in *The Strategy of Research,* both warn that there is no correlation between the difficulty of a problem and its importance. The most trivial observation can, in the mind of a scientist possessed of imagination, yield surprises of the greatest significance.

To elevate the trivial to the universal, the scientist must first of all, be a global thinker; that is, he must be able to perceive how certain principles apply to diverse phenomena. The biochemist Albert Szent-Györgyi provides a good example. His discovery of the universal principles by which oxygen reacts with living tissue stemmed from his observation that bananas and lemons react differently with oxygen: bananas turn brown when they are bruised, but lemons do not. He concluded that lemons contain something that affects the way they react with oxygen and later found that something to be ascorbic acid—Vitamin C. But Szent-Györgyi's insight did not end there. He realized that similar oxidative reactions must occur in all living organisms and went on to demonstrate how muscle tissue uses oxygen. "Looking back on this work today," he said years later, "I think that bananas, lemons, and men all have basically the same system of respiration, however different they may appear."

In the search for universal truths, a scientist is also wise to know intimately, even to identify with, the things or creatures he studies. Lorenz was fully aware of all his animals' normal behaviors—feeding, fighting, mating, nesting, imprinting, rearing, and so on—so he could recognize when a behavior was exceptional. And Pasteur and Fleming had the same complete familiarity with microbes. But intimacy means more than mere knowledge. In an interview, the geneticist Barbara McClintock, winner of the 1983 Nobel Prize in medicine, described her method of research as having "a feeling for the organism." Speaking of her work on the chromosomes of the *Neurospora* fungus, she said:

> I found that the more I worked with them, the bigger and bigger [they] got, and when I was really working with them I wasn't outside, I was down there. I was part of the system. . . . I even was able to see the internal parts of the chromosomes—actually everything was there. It surprised me because I actually felt as if . . . these were my friends. . . . As you look at these things they become part of you. And you forget yourself. The main thing about it is you forget yourself.

The mathematician Jacob Bronowski, in an essay in *Scientific American,* wrote that it is this "personal engagement" of the scientist that differentiates him from a mere technician. The physicist-philosopher Michael Polanyi calls this "personal knowledge."

The reward for such internalization of subject matter is intuition. The scientist learns to sense what is expected, to *feel* how the world ought to work. Peter Debye, a Dutch-born American who won the 1936 Nobel Prize in chemistry for his work on molecular structure, said once that he would ask himself, "What did the carbon atom *want* to do?" The virologist Jonas Salk, discoverer of the polio vaccine, writes in *Anatomy of Reality*, "I would picture myself as a virus, or as a cancer cell, for example, and try to sense what it would be like to be either. I would also imagine myself as the immune system . . . engaged in combating a virus or cancer cell."

In essence, intuition is the ability to sense an underlying order in things, and thus is related to still another mental tool that is indispensable to the working scientist: the perception of patterns, both visual and verbal. The Russian chemist Dmitri I. Mendeleyev's periodic table of elements is a classic example of how ordering facts yields new insights. Before he conceived it, in 1868, chemists had had great difficulty perceiving relationships between the elements. Mendeleyev noticed that when he arranged all the elements on a chart, according to their atomic weights, the chemically related elements appeared at regular, or periodic, intervals. (For example, magnesium, calcium, and strontium, all of which occur in the same column, have the same valence, or number of orbitals available for bonding.) His table had many gaps, but Mendeleyev correctly predicted the existence of missing elements, and scandium, gallium, and germanium, among others, were duly discovered during his lifetime.

All good theories contain, at heart, an ordering process that reveals hidden patterns. Consider how Pasteur discovered the phenomenon of molecular asymmetry—the way organic molecules exhibit what are called right-handed and left-handed forms. The discovery grew out of Pasteur's search for the molecular differences between racemic and tartaric acids, both of which are wine byproducts that form crystals on the inner surfaces of casks during the fermentation process. The German chemist Eilhard Mitscherlich had concluded that the two acids not only had the same chemical composition and specific gravity but also seemed to form identical crystal structures. The only difference between them, he believed, was that a beam of polarized light would pass directly through a racemic acid crystal but would be bent, or deflected, as it passed through a crystal of tartaric acid.

Pasteur was puzzled by the notion that two crystals, identical in structure, would differ in this key respect, for studies of quartz crystals had suggested that differences in their ability to bend polarized light always corresponded to differences in crystal form. He hypothesized that Mitscherlich had been wrong and that the light-deflecting racemic molecules would turn out to be asymmetrical in structure and the tartaric ones, symmetrical. Pasteur placed samples of both acids under the microscope and noted that there were in fact slight irregularities in the racemic molecules. Yet he also discovered, to his surprise, that the tartaric molecules were slightly irregular. This finding shot down Pasteur's initial hypothesis—that a crystal's optical activity reveals whether its molecules are symmetrical or asymmetrical—but the paradox led him to a better theory. By designing further experiments, he figured out that tartaric acid had only one asymmetrical form, a right-handed form that caused it to bend polarized light, whereas racemic acid had two asymmetrical forms—right-handed and left-handed—which nullified each other's ability to deflect light.

Pasteur's advantage over the various crystallographers who had studied the same molecules and failed to detect their asymmetry was not that he had better eyesight (he was nearsighted) or that he was better at constructing hypotheses (indeed, his first one was wrong). It was that his logic and perceptual skills (as a teenager, he had been trained as an artist) gave him an edge. First, he insisted that the tartrates fit the pattern of previously studied compounds, and, second, he looked at molecules that other scientists regarded as identical, and he recognized differences in their structures.

Verbal patterns are sometimes just as suggestive as visual ones. Fleming, you will recall, had no reason beyond the verbal symmetry of runny bottoms and runny noses to seek a biological connection. And more than one valuable idea has begun as a pun. In 1943, when the biologist Ralph Lewis, of Michigan State University, started a study of how fungus is disseminated by insects, he could not get his flies to pick up the fungal spores he was placing in their midst, so his experiment came to a halt. The problem was that the spores were drying out in the laboratory. In nature, spores are released by the fungus in a sweet, moist substance called honeydew. Free-ranging insects are attracted to the fungus by the honeydew, and it sticks to their legs, carrying the embedded spores with it. The question that occurred to Lewis was, Would honey do? In fact, honey would do just fine, and the experiment could proceed. Lewis's idea was not a formal, logical inference, but it was a perfectly good one.

It should be clear by now that scientific discovery is never entirely accidental. It holds an element of surprise, to be sure, the effective surprise that changes a person's perception of nature. But the best scientists know how to surprise themselves purposely. They master the widest range of mental tools (including but certainly not limited to, game playing, universal thinking, identification with subject matter, intuition, and pattern recognition) and identify deficiencies or inconsistencies in their understanding of the world. Finally, they are clever enough to interpret their observations in such a way as to change the perceptions of other scientists, as well. As Albert Szent-Györgyi put it, "Discovery consists of seeing what everybody has seen and thinking what nobody has thought."

We are forced to accept still another conclusion. The process of discovery is not distinguishable from that of logical testing. In each of the examples discussed above—Pasteur, Minkowski, Fleming, and all the rest—the tests neither validated nor invalidated the initial hypotheses but ended in surprise. What each of these scientists discovered was a new problem, an anomaly that led to the discovery of something else. Thus, it appears that the most important discoveries arise not from verification or disproof of preconceptions but from the unexpected results of testing them.

This has practical implications. At present, we fund experiments whose results are foreseeable rather than those that are most likely to surprise. Similarly, we train scientists almost solely in the methods of demonstration and proof. And students are evaluated on their ability to reach correct, accepted conclusions. This sort of education is necessary, but it is also insufficient, serving only to verify what we know, to build up the edifice of codified science without suggesting how to generate problems of the sort that lead to new discoveries.

A startling conclusion? Perhaps, but this is the message contained in the work of Pasteur, Minkowski, and Fleming, and dozens of other successful scientists. What is intriguing is how historians and philosophers persist in ignoring their testimony. Is it that we perceive only what we expect to see?

QUESTIONS FOR DISCUSSION

1. Why, according to Root-Bernstein, have traditional discussions of scientific method been inadequate?
2. Why are the strategies of discovery less easily codified than the rules of scientific proof?
3. What is the function of play? Why is it important? What habits of mind, attitudes, etc. does it foster—or reflect? In what ways does play contribute to the scientist's ability to "court the unexpected"?
4. The author quotes several scientists—among them Barbara McClintock and Jonas Salk—to support his view that "a scientist is also wise to know intimately, even to identify with, the things or creatures he studies." How does the scientist achieve such intimacy? Does it involve a rejection of "scientific objectivity"?
5. Compare the author's statement, "All good theories contain, at heart, an ordering process that reveals hidden patterns," with Popper's (in Part II) that "every 'good' scientific theory is a prohibition—it forbids certain things to happen."

MAHLON HOAGLAND

Preface from *Toward the Habit of Truth**

Mahlon Hoagland received his degree in medicine from Harvard University and went on to a distinguished career in molecular biology. In addition to his pioneering work as a research scientist, Hoagland served for many years as head of the Worcester Foundation for Experimental Biology. He has received numerous honors and awards, including the Franklin Medal (1976), and is a member of the National Academy of Science.

Hoagland's greatest achievement, perhaps, is his discovery of transfer RNA. In 1953, Watson and Crick had discovered the double helix structure of DNA, thereby giving us a new understanding of the basic genetic structure underlying all living things. A number of questions remained, however, among them the problem of how the genetic information contained in our genes is translated into living substance. Hoagland's discovery resolved this problem.

In the following selection, taken from his "Preface" to Toward the Habit of Truth *(1990), Hoagland discusses various aspects of science, including the nature and role of "method" in scientific practice and the importance of honesty in what is, after all, a shared and essentially communal enterprise.*

This essay touches on a number of themes taken up elsewhere in this collection: on the importance of honesty (in the broader context of scientific ethics), see Sayre (this Part); on "method," see especially Bauer (Part I); on the role of aesthetics in scientific explanation (theory), see von Baeyer (Part II).

The late distinguished biologist Sir Peter Medawar has written that what non-scientists *think* scientists do and what many learned academics *say* scientists do, and indeed what scientists themselves say they do, cannot be relied upon. To understand what science is really about, he suggests, we must eavesdrop. Thus, in any laboratory we might hear the following exchange:

"What gave you the idea of trying _____?"

"What happens if you assume that _____?"

"Actually your results can be accounted for on a quite different hypothesis."

"Obviously a great deal more work has got to be done before _____."

"I don't seem to be getting anywhere."

This kind of talk does not suggest that scientists hunt for facts or follow prescribed methods, still less that they are busy formulating laws. They are instead "building explanatory structures, *telling stories* which are scrupulously tested to see if they are stories about real life."

A scientist quickly realizes that there is no single acceptable way of getting at truth. Each explorer lays out his own itinerary into the unknown: identifies a problem, produces ideas—imagined scenarios—of the way things might actually be, and then tests the ideas by experiment. One key to success is the selection of a good system, or model, that can be manipulated and controlled to give clear answers to incisive questions. As François Jacob has written, in science one endlessly play[s] at setting up a fragment of the universe which the experiment . . . rudely correct[s]." The process is a wholly natural, if more sophisticated and arduous, extension of the one we all use from birth, intuitively, to build our picture of reality. The process is *active:* in Jacob Bronowski's words, "Science does not watch the world, it tackles it."

The questions scientists ask generally take the form of predictions: if I do this, that should happen. If a prediction is borne out by experiment, it builds confidence in the idea—but does not necessarily prove it right. The failure to affirm a prediction by experiment suggests there is something seriously wrong with the idea that generated it. Thus, good ideas often are those that suggest ways they can be proved false. If a hypothesis offers *no* way to prove itself true or false, it is not useful scientifically. It should be thrown away; it is a belief. Knowledge advances on the wings of testable ideas, not of beliefs.

The conclusions scientists arrive at are statements that have a greater or lesser probability of reflecting reality; they are never certainties. They gain strength as accumulating tests continue to verify them or prove reasonable alternatives to be false. We see not what nature actually is but the extent to which it agrees with our ideas. Thus, the possibility that some of our most cherished truths may someday turn out to be wrong can never be completely ruled out.

While scientists strive for objectivity, they are subject to the same emotions as in other endeavors: frustration, disappointment, exasperation, astonishment, delight—and lots of wishful thinking. After all, the unknown vastly exceeds the known; we start with few and often deceptive clues, and proof is devilishly hard to come by. Science inevitably is a generator of surprises. The likelihood of selecting a really "hot" problem ripe for the solving is not high, and the chances of having the right ideas for solving it are no better. For every clean, aesthetically satisfying, prediction-verifying experiment, there will be a hundred duds that require a return to the drawing board. Much of the time things go just plain wrong: ideas are not incisive enough, the apparatus breaks down, the wrong ingredients are added, the test tube cracks, and so on. Failed experiments have value, of course. They are signposts telling others where not to go; and they are a kind of personal album of the excitement, hopes, and optimism that drive the whole process.

Science obviously places a high premium on honesty. Misconduct in science in the form of deliberately faked experiments is rare. In the form of sloppiness, carelessness, bias, and the influence of wishful thinking, however, it is not so rare. As "big science" burgeons with large teams of workers, intense competitiveness, and a growing influence of the profit motive, scientists will have to watch their behavior more carefully.

There is an even more subtle form of dishonesty in science—a kind of distortion of history. Often we write up our work in such a way as to put our performance in the most favorable light; there seems to be a need to make our story appear as prescient as possible. A few scientists have, however, made the effort to "tell it like it is"—with salutary effect. A pioneer in the practice is James D. Watson. His enormously successful book *The Double Helix* (which he originally entitled *Honest Jim*) is an exceptionally candid, no-holds-barred personal account of his journey with Francis Crick to the revelation of the structure of DNA.

Scientists usually take for granted the spirit of optimism that attends modern science, a spirit firmly rooted in the phenomenal success of the enterprise. I find myself often wondering, with admiration and awe, how my scientific brethren of earlier times—hemmed in as they were by superstition, dogma, weird nontestable explanations, and the hostility of established authority and lacking today's easy access to companionship in the search—summoned the extraordinary courage to ask nature straightforward questions.

The widespread cultural faith that nature is comprehensible, that the unknown *is* knowable, is relatively new in human history, going back not much more than three hundred years, to Isaac Newton. Consider the terror that nature in the raw must have caused our ancestors and how fiercely over the centuries free inquiry has been feared and resisted by religious authority and popular superstition. I marvel at the audacity of the lone scientist searching for answers under the eye of an omnipotent God reputed to be easily able to keep the truth from anyone.

Sensitivity to one's own ignorance is an essential state of mind for the scientist, and highly recommended for the nonscientist as well. We honor those explorers of the past when we practice healthy skepticism. We are not likely to ask serious questions or make meaningful predictions without first assessing what we do not know. Belief is the easy way; it "explains" everything by explaining nothing. My colleague Rollin Hotchkiss remembers as a boy bringing a clock to his father and asking him why it had stopped. His father removed the back, and a fly fell out on the table. Said his father, "Well, of course, the engineer's dead!" The dead engineer is still with us in the facile nonexplanations used by such pseudosciences as creationism, astrology, occult and so-called psi phenomena, scientism, and so on and on. Still, we have come a long way.

The pleasure in the exercise of skill to transform idea to substance is common to art and science. My lifetime habit of doing sculpture alongside science has convinced me that the common enjoyment is indeed in the action: manipulating material and technique to uncover simplicity, order, and harmony below the surface of things. In sculpture, a form of one's imagining emerges from the density and obscurity of a material, just as in science a new vision of reality arises from mystery by the action of experimentation.

It is easy to oversentimentalize the similarities of art and science. In science, an idea can become substance only if it fits into a dynamic accumulating body of knowledge—a progression of understanding. Each new piece of work is subject to validation in respect to its compatibility with the bigger picture. It is inspected, tested, tentatively accepted, modified, perhaps discarded. There is really no equivalent progress or cumulativeness in art. Art may progress in technique or indebtedness to predecessor but not in content. In art, creator and created content are inseparable.

In science, however, the discovery is uniquely the discoverer's only in terms of priority and in the way it was made. Its content could have—would have—been found by others. If Columbus had not made the trip, someone else would have; indeed, he was not the first anyway. If Newton, Darwin, and Einstein had not been around, others would have come up with insights that in the end would have built the same edifice of knowledge. Only the history of the construction process would be different. Furthermore, science is rich in examples of simultaneous, independent discoveries, such as Darwin's and Wallace's independently attained visions of natural selection.

Gunther Stent has argued that scientific discoveries are as uniquely a scientist's as works of art are an artist's. He quotes Medawar, for example: "The great thing about [Watson and Crick's] discovery [of the structure of DNA] was its completeness, its air of finality. If Watson and Crick had been seen groping toward an answer, . . . if the solution had come out piecemeal instead of in a blaze of understanding: then it would still have been a great episode in biological history, but something more in the common run of things; something splendidly well done, but not done in the grand romantic manner" that it, in fact, was. Stent also holds that two discoveries arrived at simultaneously are rarely, if ever, identical and therefore are uniquely each discoverer's.

It seems to me that here Stent confuses content with manner of discovery. Of course, every scientific discovery is unique in the *way* it is made. Medawar's comment supports the view that there is little unique about Watson and Crick's relation to their contribution beyond priority and the style in which it was made—which was unquestionably magnificent. And as far as simultaneous discoveries are concerned, their similarities are far more significant than their differences. Wallace may not have had the historical impact that Darwin did, but he independently came up with essentially the same germinal idea—the same content—that provided all of biology with a unifying theme.

In the march of scientific discovery, then, its artisans blend into history like the builders of the great cathedrals. Scientists must be pretty high on egotism to avoid acknowledging their own expendability. This reality, together with teaching us how little we know and how difficult what we do know was to come by, makes science a profoundly humbling experience.

As Max Delbrück, the physicist turned biologist who is generally considered the inspired founder of molecular biology, said on the occasion of receiving the Nobel Prize,

> The books of the great scientists are gathering dust on the shelves of learned libraries. And rightly so. The scientist addresses an infinitesimal audience of fellow composers. His message is not devoid of universality but its universality is disembodied and anonymous. While the artist's communication is linked forever with its original form, that of the scientist is modified, amplified, fused with the ideas and results of others,

> and melts into the stream of knowledge and ideas which forms our culture. The scientist has in common with the artist only this: that he can find no better retreat from the world than his work and also no stronger link with the world than his work.

Once on a summer evening, I stood with a friend watching fireflies light up a field. Moved by the beauty of this commonplace event, I said, "You know, we now know what happens in the firefly's tail to make that light." Expecting a sign of curiosity, I got instead, "I enjoy beauty as it is—don't spoil it by explaining it!"

As children, we all possess a natural, uninhibited curiosity, a hunger for explanation, which seems to die slowly as we age—suppressed, I suppose, by the high value we place on conformity and by the need not to appear ignorant. It betokens a conviction that somehow science is innately incomprehensible. It precludes reaching deeper, thereby denying the profound truth that understanding enriches experience, that explanation vastly enhances the beauty of the natural world in the eye of the beholder.

When I discovered the initial step in the synthesis of proteins, I was bowled over by the beauty of the process by which all living things use energy for the construction of their substance. Soon after that discovery, William McElroy of Johns Hopkins University, in Baltimore, who was investigating the nature of the reactions that produce the firefly's light, found that the initial step in that reaction was identical to the one I found for protein construction. The same first step for building living substance and for bioillumination—an astonishing connection!

Again, when Paul Zamecnik and I discovered transfer RNA, it emerged as the *physical connection* between DNA, the source of all genetic information, and the machinery that translated that information into a new living being. That Francis Crick had imagined the existence of the connection in advance of our discovery was sheer poetry.

It is often the scientist's experience that he senses the nearness of truth when such connections are envisioned. A connection is a step toward simplification, unification. Simplicity is indeed often the sign of truth and a criterion of beauty. Scientific truth may be complex: the protein synthesis machinery of cells involves the interplay of hundreds of separate proteins and other complex molecules. But when these molecules become meaningfully interrelated in a mechanism serving all of life, they take on the simple beauty of explanatory law. As Jacob Bronowski said, "It is the sweeping simplicity of [nature's] means that overwhelms [the scientist] with a sense of awe. This is what makes nature beautiful . . . the simplicity of the materials which make so many patterns, the unity under the surface chaos. Unity is the scientist's definition of beauty, and it makes nature beautiful to him all his life."

The aesthetics of explanation in biology—the view that knowledge of underlying mechanisms of living processes enhances their beauty—is most dramatically and grandly brought home by the study of evolution, the theme that has woven its way into the fabric of all biology. Evolution unifies the study of life by showing continuity, connection, and theme among the array of forms and functions in the living world.

When I started out in research, it was still not thought properly scientific, at least among biochemists, to ask *why*—why living creatures are the way they are, why they do the things they do. Such questions smacked too much of metaphysics, wandered too far beyond the bounds of objectivity. One was wise to stick to *what* and *how*. This strictly analytical approach to biochemical problems, the dissection of living systems to reveal fundamentals of structure and mechanism, has been, and continues to be, prodigiously successful.

It has revealed, for example, that most of the basic elements of structure and function of all organisms, from bacteria to humans, are remarkably similar—in some cases, identical. We all use the same sorts of proteins made up of an identical set of twenty amino acids; we all use the same nucleic acids made up of the same four bases as genetic material. We all have similar machinery for oxidizing our food and producing our energy and for doing our cellular work, including the building of ourselves. We store, replicate, and use genetic information in the same way. The genetic code, the cipher for translating inherited information into living substance, is the same in all of us. These truths are pillars of support for evolution's first premise—that we all had a common origin.

Darwin's imaginative leap linking a common origin, hereditary variation, environmental selection, and relative reproductive success was the grand justification for asking a whole new set of questions. It was a framework upon which we hung our understanding of how the great variety of living forms has arisen since the earliest bacteria appeared some three and a half billion years ago. The science of genetics came of age seventy years after Darwin to begin the process of providing an understanding of the mechanisms underlying hereditary variation.

Before the 1940s, bacteria, by far the oldest and most abundant forms of life on our planet, were thought to have unique capabilities that accounted for their variation and environmental adaptability. This was largely because they seemed very different from higher forms: they did everything much faster, and we knew very little about their genetics. Then, in the 1940s, it was learned that the inheritance of bacterial traits could be altered by exposing living bacteria to materials extracted from dead bacteria, material that later was identified as the chemical DNA. Genes, once considered nebulous "units" of inheritance, were soon discovered to be stretches of prodigiously long DNA chains, each gene responsible for making one protein molecule. Furthermore, it was unequivocally established that the variation and adaptation of bacteria resulted from the environmental selection of rare mutant forms, as in all other forms of life. Later, when bacteria were found to conjugate, to mate sexually and thereby recombine genetic material by mechanisms fundamentally similar in all living forms, they were seen to be part of the living continuum. Their simplicity and rapid growth made them ideal experimental models. One could observe evolution in the test tube on a shortened time scale. Since bacteria were haploid, meaning that during the major part of their reproductive life cycle their genes existed in single copies, rather than in duplicate as in most other life forms, there were no problems related to hidden, or recessive, traits. Their genetics could therefore be much more easily and precisely analyzed.

With bacteria as models, and with growing sophistication in the study of cultured cells from higher organisms, biochemistry and molecular biology proceeded to reveal the physical and chemical bases of hereditary and variation in minute detail. The revelation of the structure and mode of replication of DNA, and the mechanism by which its information is translated into protein molecules, made Lamarck's theory that acquired characteristics could be passed along to offspring highly improbable.

The more we learned, the more apparent it became that variation, obvious at the organismic level, was even more extensive at the molecular level. Variation, seen as change in an organism's protein composition upon which forces in its environment can act, was shown to be due to specific chemical alterations of information in DNA molecules that resulted in precisely predictable alterations in protein structure. Such changes in DNA were found to be of great variety, including mechanisms for substantially expanding the number of an organism's genes and for passing genes from one organism to another, and they could thus account not only for change but also for a gradual increase in complexity of organisms.

So, a century after Darwin had laid out his theory of natural selection, the molecular mechanisms needed to account for the gradual evolutionary changes in organisms and for the appearance of new species were revealed. Even more directly, molecular genetics showed that as organisms diverge in form and function, they similarly diverge in the sequences of their DNA and protein molecules. Indeed, these chemical criteria are much more accurate measures of divergence and relatedness than are anatomical criteria. For example, recent evidence that the DNA of humans and of chimpanzees is 99 percent identical suggests that many physical differences between organisms may be due to changes in genes regulating the expression of other genes, rather than to differences in the genes themselves.

The story of man's rapidly expanding understanding of his origins, beginning with inspired leaps from anatomy and paleontology through genetics to the physicochemical foundations of nucleic acid and protein structure, is one of the greatest sagas of human creativity—of the use of imagination in the rigorous pursuit of evidence to bring connection, continuity, and community to a bewildering mass of biological data. Explanation has brought harmony, pattern, meaning, and beauty in its wake. It speaks for the first time in all of man's metaphysical wanderings, directly and without maudlin religiosity or intellectual hypocrisy, to what we really are, where we came from, and how we relate to our fellow creatures.

QUESTIONS FOR DISCUSSION

1. What does the author mean when he says that "in science . . . the discovery is uniquely the discoverer's only in terms of priority and in the way it was made"?
2. How can Hoagland be so confident that "if Newton, Darwin, and Einstein had not been around, others would have come up with insights that in the end would have built the same edifice of knowledge"?
3. Does the author illustrate his point that in science "each new piece of work is subject to validation in respect to its compatibility with the bigger picture"?
4. What does the author mean by "the aesthetics of explanation"? Is this notion applicable to physics as well as to biology? Why?
5. In what way(s) does theory promote discovery?

ANNE SAYRE

The Making of a Discovery*

A widely published author, Anne Sayre was educated at Middlebury and Radcliffe Colleges and at New York University. In the early 1970's, as a Visiting Scholar at Lucy Cavendish College, Cambridge Uni-

*From *Rosalind Franklin & DNA* ©1975 by Anne Sayre. Reprinted by permission of W. W. Norton & Company, Inc.

versity, she did much of the research for her study of Rosalind Franklin's contributions to the discovery of the double-helix structure of DNA. In 1975, she published this work as Rosalind Franklin & DNA.

Sayre's study of Franklin has been highly praised. For example, Linus Pauling, himself a major player in the events that led to this fundamental discovery, described Sayre's account as "well written and illuminating." Another critic, writing in the Quarterly Review of Biology, *said that "anyone who has read, or is to read, [James] Watson's version of these events [in* The Double Helix, *1968] must also read . . . Sayre's book."*

In the following excerpt from her book, Sayre discusses the inherent contradiction at the heart of science between competition and communication/collaboration, and probes the ethical issues this contradiction gave rise to in the search for the structure of DNA. This essay can be usefully read in conjunction with Hoagland (this Part) and, with respect to its discussion of a female scientist whose achievements for years went largely unacknowledged, can be placed alongside Keller's essay on Barbara McClintock (this Part).

What was going on between 1951 and 1953 was a race, with the discovery of the structure of DNA as both the goal and the prize. This sort of contest is far from unfamiliar in science, which is a competitive business; what is odd about this instance is that some of the contestants did not know that they were racing. Rosalind was one of them.

The nature of this race is less well-documented than might be expected, much less than is usual. Modern science keeps close track of its own progress, recording itself in those innumerable journals in which scientists publish their findings: experiments, discoveries, conclusions, perceptions, and, occasionally, even their failures, lack of conclusions, bewilderments. These journals do more than simply record history. Any field of science is an intensely, and often a very rapidly, cumulative subject, the substance of which is constantly changing and expanding; and these changes in substance are absolute. In other words, any major new published discovery has an instant effect that is profound, that may require work in progress to be abandoned because it has been shown to be based upon a wrong guess or a disproved assumption, that opens out immediately prospects of new investigations, that in no case may be ignored. There is no element of choice in this either. A doctor who keeps up with the medical literature is not obliged to incorporate into his practice everything he has read about. Provided that most of his patients survive, neither the success nor the value of his work will be judged by how up-to-the-minute or innovative his methods are. But those who do research lack the option to be old-fashioned; the success or value of what they do will, indeed, be closely and invariably linked to whatever the newest relevant information on the subject may be. The published literature of science is, in fact, the subject itself, history in the retrospective sense, but also the expression of the current sum of knowledge at any given moment.

Because of this, most discoveries emerge, not out of the blue, but out of a clear line of progression, traceable in the literature. It is usually possible to perceive with some accuracy who contributed what, even to a solution that took a long time in the finding. Though opinions may differ as to the relative importance of various contributions, there is rarely any doubt of their existence. But when we come to DNA, things are not quite like this. The literature does not tell everything, and we know that it does not, because J. D. Watson wrote *The Double Helix* in order to tell us so.

We learn from *The Double Helix*, for instance, that a race was going on. Indeed, two races are mentioned, though one of them—the one against the devilishly clever brain of Linus Pauling—was apparently rather illusory; the other, however, was real enough. King's College (London) was pitted against Cambridge, or much more accurately, Rosalind and Wilkins were vying with Watson and Crick.

It was this of which Rosalind was not entirely aware. She was not a fool, of course. She did not live in an ivory tower either; she was a practicing scientist, and an ambitious one, and she was perfectly conscious of living in a competitive world. It is true that before the structure of DNA was worked out, there was no way for anyone to know that it would prove to be of almost unique importance; but certainly it was understood that the problem was a significant one, and Rosalind did not for a moment imagine that this understanding was confined to King's College. Had she, however, chosen to make out a list of rivals, it is probable that Watson and Crick would not have been on it. And this is a very unusual situation.

Scientists are communicative people; they are obliged to be. The obligation also exists in the sense of a duty requiring them to publish their findings. In science, even more than elsewhere, to suppress a truth is to consent to a lie. But science is also competitive; and because of this it is burdened with a problem that is not easily solved. If half the motive behind the duty to publish one's findings is a duty toward oneself, the other half is an acknowledgment of the necessity of pooling information and knowledge for the sake of science itself. Indeed, if this were not done, and new truths were kept secret, the progress of research would slow to a crawl, if only because time and energy and the resources of intellect would be devoted to repeating what has already been done, rediscovering what has been discovered, duplicating what already exists.

Scientists are immensely sensitive to their urgent need for free communication; for this reason, there have been many of them who have objected to the restrictions placed on exchange of information by governments which like to keep some areas of research veiled. There are few who are not uncomfortable, to say the least, when required to compromise their right to publish which is also their obligation. On the other hand, the duty to keep telling, to keep providing new truths for the benefit of others, is an unnatural one in terms of a highly competitive society. It does not exist in commerce because commerce is pure competition. A new engine invented by Ford will not be confided to General Motors; there is no moral obligation which urges that it should be; and, if General Motors has wasted a great deal of time and labor and money producing cars which cannot compete with Ford's new line and the value of its shares consequently declines, that is the nature of economic life in a capitalist and competitive society. Nobody complains.

In commerce, the motive is happily single: to make a profit, if possible. But it is not so simple for scientists. Science, after all, is not simple. As a social activity—divorced for the moment from its technical content—it is humane in the sense of aiming to provide useful knowledge for the general benefit, and not for profit. But this end is reached through means which involve as much competition as commerce does. Scientists too must eat; how well and how often they eat depends to some degree upon their making a good showing in one race or another. More than that, scientists, like most other people, have egos which need to be satisfied, which rejoice in praise and reverence, which cherish secret dreams of immortality—Newton's Law, Einstein's Theory, Darwinism, why not add another name to the list of the unforgettable?

The urge to compete and the need to communicate are in opposition, and the balance which any scientist must maintain between the two is a delicate one. Because scientists are dependent upon communication—and not only the printed word; on their own topics they tend to be obsessive talkers—they need definitions sufficiently agreed upon so that the communicating can take place in an atmosphere of some trust. This is particularly necessary because the most interesting and valuable communications are those with people who are at least potentially competitors, who are concerned with the same problems, who share the same body of knowledge. What must be known is the extent to which the person talked to is really a rival; what must be understood is the extent to which anything divulged, or developed, or discovered in the course of the communication will be appropriately credited.

Such agreement is not thrashed out anew at the beginning of every conversation; it does not need to be. It is assumed, because there is a body of practice, etiquette, manners, which is generally subscribed to, and which covers most cases. It is based upon the respect for priority of publication which governs the rewards science gives to her own—a respect which is not exactly copyright or patent, but which all the same confers a kind of right. Those named as discoverers are those who published first; on priority of publication rests the right to acclaim. Because Mendel published his discoveries very obscurely, they were rediscovered rather than communicated; but though Correns and de Vries reached the same conclusions that Mendel did, and quite independently, the credit for originality is not theirs, but his. This looks like a nicety, considering the circumstances, considering that Mendel was a long time dead and that his researches had been uninfluential, considering that those who rediscovered his work did so in ignorance of what had gone before. But it is more than a nicety; it is, indeed, the moral equivalent of copyright or patent; and if it is only moral, that does not lessen the claim, for it is not in any case ever carelessly to be assumed that either the desire to be credited with one's accomplishments or the right to that credit is always tied to profit.

Because there is no profit, no legal claim, and because the output of scientists is not only freely borrowable, but is intended to be borrowed, these scientists must either trust each other considerably, or else maintain so discreet a silence that the progress of research is impeded. On the whole they trust each other, and are right to do so. Candor is the chastity of scientists, and generally, it is diligently preserved.

One occasion for candor is the frank announcement that one has entered a race. This announcement was never made in Cambridge, nor received in London. There are circumstances which explain this; whether, as things developed, they entirely justify it is another question. There existed what might be called an administrative understanding that placed the DNA problem in Randall's laboratory at King's. Such arrangements are not ideal. Ideally, all problems should be available to all comers on an open, competitive market; and so most of them are. People in Dallas, Liverpool, Berlin, and Peking may all be grappling with the same puzzle at the same moment, and probably this is exactly what is happening. Rivals may well be operating out of laboratories in Berkeley and Cambridge, Massachusetts; and so they should be. Rivalry is stimulating and useful, and this is the way in which science works.

At least, this is the way science works most of the time. But the free-market approach is an expensive one which produces a considerable amount of duplicated effort. Therefore, there are sometimes situations in which efficiency suggests the avoidance of competition. If there are three problems urgently in need of investigation, then it is more sensible to turn over one to each of three laboratories, rather than having all the laboratories turning their attention to the most attractive problem, while the other problems languish. This sounds very good on paper; where the resources which finance science are limited, it is an appealing approach; and in more than one place and at more than one time it has been used, sometimes by formal agreement and sometimes by informal understandings.

This is an approach which scientists themselves dislike, and they tend to resist it, except under extraordinary conditions, such as during a war, for instance, when priorities intervene which need not have much to do with science itself. And they are probably wise to resist it, for research is too creative a business to profit from being narrowly channeled. But sometimes it happens. One reason why those at King's College did not realize that people at Cambridge were working on the structure of DNA was simply because an understanding existed that they weren't.

This understanding was acknowledged. For that we have Watson's word. He points out in *The Double Helix* that molecular work on DNA was, in 1951, essentially the property of King's College, and in particular, of Maurice Wilkins, and deplores the coziness of England which made it awkward for one scientist—in this case Francis Crick—to move in upon territory claimed by another: Wilkins. Though this might have been acceptable with respect to a foreigner, "the English sense of fair play would not allow Francis to move in on Maurice's problem. . . . In England . . . it simply would not look right."

The statement Watson makes may represent a partial misunderstanding of the state of affairs, for as he puts the case, it is not very convincing. He adds that it would not be reasonable to expect someone at Berkeley to ignore an important problem simply because someone at Cal Tech had started on it first; but then, one does not really expect this in England either. English scientists are no less scientists than those in other countries; they are no less ambitious; and whatever they may have learned on the playing fields of Eton or Manchester Grammar School does not leave them so obsessed with notions of fair play that they hover outside the laboratory door, incessantly murmuring, "After you, old chap."

This would be absurd; as Watson has put it, it is absurd. No reasonable person would consent to such nonsense; and this unreasonableness may be what Watson is suggesting. But England, in 1951, had not vast sums available for the financing of research, unlike the United States, which may have been the image Watson had in mind. Not so much a notion of politeness as a need to make the money go around had produced a certain amount of that resentable practice of dividing up the problems, on an informal level, evidently, and not by any bureaucratic dictation. Those who head laboratories, and want funds for which they must often apply to government agencies, are well-advised to base their applications upon a list of projects not also going on elsewhere; otherwise they may be asked to explain exactly why it is that their organization should be provided with support in order to do the same thing as someone else.

In these circumstances, it is not unusual, and it is sensible enough for those who head laboratories to communicate with each other and to reach some agreements intended to avoid duplication. Such an agreement—unenforceable, informal, very possibly in the noblest scientific sense undesirable—existed between the administration of the laboratory at King's and its opposite numbers at Cambridge. The reason for making it was neither stupid nor absurdly punctilious. But it was an embarrassment. The

Cavendish Laboratory at Cambridge had its problems to work on; the biophysics laboratory at King's College had others; and what King's had included DNA. To announce openly that a race for the DNA prize had now begun between the two institutions would have violated an agreement—admittedly, a gentleman's one—and not to announce it was contrary to that complex understanding on the basis of which scientists can manage to be both rivals and trusting friends.

These were the horns of a very real dilemma. But, then, the race was not really between the two institutions. Sir Lawrence Bragg, who directed the Cavendish, was by no means inclined to encourage the investigation of DNA under his roof. On the contrary, according to Watson, what Bragg said was that he and Crick must give up DNA, and apparently—again, according to Watson—Bragg held firmly to this decision until the work that was going on anyway, in spite of his fiat, had reached a point at which no scientist could advocate, or even tolerate, its suppression. The race was, then, a personal one, between Crick and Watson on the one hand, and King's College on the other. This is a fact of great importance.

No one at King's College seems to have realized what was going on. That there were people in Cambridge—and in a number of other places—who were interested in DNA, yes; that they were developing their own ideas and techniques, yes. But that there was a hell-for-leather gallop for the finish line, toward the independent publication of a discovery, no. Rosalind did not know, but she was not in contact with Crick at all during this time, and not much in contact with Watson, whom she disliked. Wilkins appears not to have known, although he saw a good deal of Watson and something of Crick—who was in any case an old acquaintance—and his relations with both were amicable. Randall certainly did not know. Ought they, individually or as a group, to have suspected? Hindsight says yes; but at the time such suspicions would have seemed faintly paranoid, when, after all, it was understood . . .

No one suspected. Whether this ignorance of the true state of affairs made the least difference is an arguable point. To know that other people are scrambling after the same prize can be something of a spur, but only within limits. Rosalind, for example, was so much in the habit of pushing herself to the full extent of her energy and application that competition could not have elicited a great deal more than was already forthcoming. Nor was the situation one in which recognition of rivalry might usefully have produced something more in the way of money or support, because neither was required. What awareness might have produced was, perhaps, discreet silence, or possibly—but this is pure speculation, of course—some lessening of the friction between Rosalind and Wilkins by providing a common cause in which they might have been able to join sufficiently to overlook what each found to be failings in the other. Nobody can say that the conversations they might have had, had they been able to converse at all, would have been significant and productive. All that can be said is that neither felt any external pressure that required them to try.

Largely because of this situation, although Rosalind came very close between January 1951 and March 1953 to solving the DNA problem, she was beaten in the end by Crick and Watson, who had in their success more help from her work than she ever knew they had received.

How can this be? Science is endlessly communicative. Scientists publish, and what they publish is public property. Anyone can seize upon an idea, a method, a result, a perception, a theory, appropriate it to his own purposes, and carry on from there. All that is required is the small, proper note of acknowledgment which makes perfectly clear who contributed what. Scientists also talk. They give lectures and seminars; they have conversations. When they write, however, what they write is not always for publication—apart from correspondence, there may be reports on work in progress not intended for immediate circulation, containing material, for instance, not ready for public scrutiny and criticism, or perhaps the raw data from which a publication will one day emerge. (And it is important that there should be no pressures leading to premature publication merely for the sake of claiming the problem.)

Scientists write; they talk; and when they talk, what they say is sometimes on the record, sometimes it is not. No professor would dare to lecture on his unfinished or unpublished work if he had to assume that some listener would transform his teaching into a paper that would secure to someone else the credit for the notion or the result. No two people working in the same field could enjoy shoptalk, a chat, a good argument, if each did not feel bound by the same limitations. Talk is essential, talk stimulates, arguments clarify, speculations which are thrown out to the winds may fall like seed to spring up with a crop of perceptions.

That very few scientists are really taciturn in the presence of others in their special field is understandable, they cannot afford to be. And they do not need to be. This inveterate chattiness is no threat, because it is carried on within limits prescribed by convention, on the understanding that if brains are being picked, then a proper and sufficient acknowledgment of this will turn up in due course. Brain-picking can be so mutual, indeed, that a good many fruitful collaborations have been unplanned, being no more than the result of one person's musing to another, who muses back, until out of this meeting of two minds a joint work is produced. More frequently what occurs is an exchange of profitable suggestions, usually confessed to in that last paragraph of published papers where the acknowledgments lie. That the confession is made is, of course, a matter of etiquette; it is unenforceable, except in the area of reputation. But let it not be thought that good manners are unimportant. When we apply to them the name of ethics, we are not elevating right behavior to a higher plane than it deserves, but only recognizing how essential to community survival right behavior is.

Circumstances which discourage right behavior are to be deplored. The dividing up of intellectual problems into private preserves not to be poached upon may well be a deplorable practice because it discourages free intellectual activity on the one hand and on the other discourages candor when intellectual activity has asserted its freedom. Two people taking a great "unofficial" interest in a problem "officially" not theirs to attack can, no doubt, find candor hard to achieve. It appears that in the case of DNA something close to this occurred. There is reason to suppose that candor will never be achieved in this case. The record has, quite simply, become extremely unclear. There is some mystery where there ought to be history. There is insufficient documentary evidence. Too many of the events which led up to Crick's and Watson's publication of their monumental discovery can be reconstructed only from reports of what people think they remember; these reports conflict, and sometimes conflict enormously. This is not to be wondered at. Memory is never very reliable, and at a remove of twenty years or so from what it is recalling, it is rarely to be sworn to; this is true even when what it is dredging up out of the past is neutral material, not calculated to trouble anyone, and this is not a neutral instance.

The conflicting reports indicate a lack of neutrality; sides have been taken, and however multiple and shifting these may be, loyalty no doubt forbids dispassion. It is not—and this should be clear—that anyone can be said to be lying. When an atmosphere grows thick enough with justifications, explanations, rationalizations, postures, and regrets, not to omit occasional hostilities, untruth disappears just as surely as truth does. What remains is a series of viewpoints, none of which can be taken as proof of anything more than the state of mind of whoever expresses them, always speaking sincerely enough, beyond question, but not precisely with the cool calm voice of historical accuracy.

Much is lost, and this is a pity. It is not less regrettable because it is not a unique instance. Fact has been devoured by opinion before this. That is why we do not know much about Richard III, except for rumor. Lost facts are not always replaceable; certainly opinion does not replace them, nor does a legend. Nobody involved in either side of the DNA work kept a journal and, though the existence of a contemporary written record is often rumored, none of the people to whom this diary is attributed has ever admitted to its existence; certainly no one has produced it. The only approximation of a running account is in a set of notebooks which Rosalind kept, meticulously entering the course of her experiments. But these are not enough—they are a scientist's daily logbooks, testifying to a neat and precise turn of mind, but containing no personal or unnecessary comments whatever.

Among all the opinions which exist—and surely this is odd—Rosalind is the one person involved in the events whose point of view cannot be reported, for the overwhelmingly simple reason that she never had one. The events which have subsequently been the subject of so much curiosity, so much speculation, which have been so much explained and justified, were ones of which she was unaware. She was, in fact, profoundly innocent, she never asked, never guessed, never was told.

QUESTIONS FOR DISCUSSION

1. Why are scientists obliged to publish?
2. What paradox at the very heart of science does this obligation give rise to?
3. Do you agree with Sayre that "In science, even more than elsewhere, to suppress a truth is to consent to a lie"?
4. What does the author mean when she says that "candor is the chastity of scientists"?

LUIS ALVAREZ

Fission*

Educated at the University of Chicago, Luis Alvarez was for many years a professor of physics at the University of California, Berkeley. During his distinguished career, Alvarez received numerous awards, including the Albert Einstein Award (1961) and the National Medal of Science (1964). In 1968 he was awarded the Nobel Prize for his contributions to elementary particle physics. He died in 1988.

*Originally appeared in *Alvarez: Adventures of a Physicist*, ©1987 by Luis W. Alvarez. Reprinted by permission of BasicBooks, a division of HarperCollins Publisher, Inc.

Alvarez's scientific achievements were both practical and theoretical. Early in World War II, he developed three important radar technologies, including the microwave system that allows aircraft to land safely in bad weather. Later on at Los Alamos, he worked on the development of the atomic bomb with Enrico Fermi, J. Robert Oppenheimer, Edward Teller and others. Alvarez's brilliant theoretical work was perhaps most clearly illustrated in his study of high-energy particles through the use of a hydrogen bubble chamber. By the early 1960s, this work had resulted in a dramatic increase in the number of known particles. By combining the hydrogen bubble chamber with new methods of data analysis, Alvarez opened up entirely new possibilities in particle physics.

However, Alvarez's curiosity extended well beyond traditional physics. Indeed, he was considered by some to be a "wild idea man." His wide-ranging interests included the use of cosmic rays to determine the structure of King Kefren's Pyramid at Giza and the proposal (with Walter Alvarez, his geologist son) of a radical (and controversial) theory to explain the sudden disappearance of dinosaurs 65 million years ago.

In the following essay, taken from Alvarez: Adventures of a Physicist, *the author recounts from a personal perspective his and his colleagues' near misses with respect to the discovery of nuclear fission and other breakthroughs in atomic physics. His account suggests something not only about the nature of discovery but about science itself: that although discovery often involves chance (combined, of course, with individual insight), it always occurs within a context created by the work of others. In this sense, even the most original of discoveries can be seen as arising out of a broader communal effort. On this point, see the essays by Hoagland and Sayre in this Part.*

I learned about the discovery of nuclear fission in the Berkeley campus barbershop one morning in late January 1939, while my hair was being cut. Buried on an inside page of the *San Francisco Chronicle* was a story from Washington reporting Niels Bohr's announcement that German chemists had split the uranium atom by bombarding it with neutrons. I stopped the barber in mid-snip and ran all the way to the Radiation Laboratory to spread the word. The first person I saw was my graduate student Phil Abelson. I knew the news would shock him. "I have something terribly important to tell you," I said. "I think you should lie down on the table." Phil sensed my seriousness and complied. I told him what I had read. He was stunned; he realized immediately, as I had before, that he was within days of making the same discovery himself.

In their classic series of papers reporting artificial radioactivity induced in a large number of chemical elements by neutron bombardment, Enrico Fermi and his colleagues at the Physics Institute in Rome in 1934 had noted that bombarding uranium with neutrons gave rise to a variety of radioactivities of different half-lives. They suspected, and tried to prove, that among those artificially created radioactivities were new elements beyond uranium (element 92 in the periodic table), transuranics never before seen on earth. "These two [chemical] reactions," they noted near the end of a paper mailed in July 1934, "appear to confirm the hypothesis that we have elements of atomic number higher than 92."

The Fermi group's proof involved chemical studies designed to show that the newly created radioactivities in uranium could not be attributed to any "nearby" chemical element. They compared the radioisotopes with elements of lesser atomic number, all the way down the periodic table to lead, atomic number 82, to which uranium eventually decays. Soon afterward a German chemist, Ida Noddack, published a critical demurrer: Fermi could not claim the discovery of new transuranium elements, she argued, until his unidentified radioisotopes had been compared with every element in the periodic table.

No one took Noddack seriously. The notion that uranium could turn into a lighter element in the middle of the periodic table under bombardment by nothing more energetic than thermal neutrons was self-evidently ridiculous; to do so, it would have to split, and the nucleus, we thought then, before Bohr elaborated the liquid-drop theory, was harder than the hardest rock, bound together by powerful forces—powerful enough to resist the electrical repulsions of all the protons. Everyone knew that the alpha particle—a helium nucleus, atomic number 4—was the largest chunk of nuclear material that could be chipped out of an atom. Nor was Noddack an entirely credible critic. She had shown her ability as a chemist by codiscovering, with her husband, the element rhenium, but the Noddacks had later announced the discovery of another element that proved to be mistaken, and they had continued rather shabbily to insist on the correctness of their work when the evidence demonstrated otherwise.

I was bothered at the time that the Fermi transuranics didn't fit the pattern of other radioactive elements. Instead of decaying, as we said, "downhill to the floor of the valley of stability," they decayed uphill, into a region that ought to be progressively more unstable. I had long been responsible for maintaining the big isotope chart that hung on the wall of the cyclotron control room, and every time I looked at it I was affronted to see the so-called transuranium elements decaying in the wrong direction. I knew something was off-key, as did everyone familiar with nuclear theory, but the correct explanation entailed such a radical departure from contemporary understanding that no one pursued the matter. It's a shame that Frau Noddack didn't follow up her own suggestion. She might have made, three years earlier, the epochal discovery Otto Hahn and Fritz Strassmann made at the Kaiser Wilhelm Institute for Chemistry in the Berlin suburb of Dahlem in 1938.

I had spent several evenings before then following my father's advice, trying to think of a way out of the madness I observed on the isotope chart. I would still feel like an idiot if Enrico Fermi, who was infinitely smarter than I was in such matters, had not also resisted Noddack's lead. Fermi received the Nobel Prize in 1938, only weeks before the Christmastime discovery of nuclear fission, "for his demonstrations of the existence of new radioactive elements produced by neutron irradiation and for his related discovery of nuclear reactions brought about by slow neutrons." He could have been awarded the prize for any of a dozen different theoretical or experimental discoveries in the course of his richly creative scientific life; it's ironic that the first citation on his Nobel diploma happens to record the only scientific mistake he made that I know of.

Phil Abelson was among my best graduate students. He had been trained in chemistry as an undergraduate, and he wondered about the chemistry of Fermi's transuranics. Since I had recently demonstrated that many radioisotopes emitted characteristic X-rays, Phil thought he might be able to determine the atomic number of the transuranium radioisotopes by measuring their X-ray spectra.

From the multitude of radioactivities produced by the bombardment of uranium with slow neutrons, he chose a radioisotope with a three-day half-life and separated it chemically from its cohorts. He showed that it gave off X-rays that were absorbed in aluminum the way the L X-rays of a transuranium element should be. He prepared to record its X-ray spectrum on a photographic plate. By measuring the position of the diffracted X-ray spectral lines, he could assign an atomic number to the isotope and determine if it was in fact what Fermi's group had proposed.

He was unlucky in the numerical values involved; on his first attempt the diffracted X-ray lines didn't hit his photographic plate. During the next week he would have changed his observation angle and obtained the telling pictures. They would have shown the simple K X-ray lines of a light element rather than the complex L X-ray lines of a heavier element, and the story of how Philip Abelson discovered fission would be history. But I bolted from the barber chair, and Phil wasn't given that extra week. Once he knew of fission, he quickly found that he was looking at iodine K X-rays; his isotope was tellurium, atomic number 52. With Phil hovering over me, I wrote out a telegram to the *Physical Review* for him, which he signed. The report "Cleavage of the Uranium Nucleus" appeared in the same issue with several other verifications of the fission discovery, one of them by Kenneth Green and me. Phil went on to a distinguished career, contributed vitally to the production of uranium 235 for the first atomic bomb, and served for many years as the much-respected editor of *Science.* His last work at Berkeley was as codiscoverer of neptunium, the first of the transuranium elements. His partner, Ed McMillan, won the Nobel Prize for that work, but, for reasons I never understood, Phil missed out on that high honor.

I also narrowly missed discovering fission. Like many of my colleagues around the world, I looked for long-range alpha particles coming from uranium bombarded by slow neutrons. Also like them—including Fermi's group—I covered the uranium with just enough aluminum foil to block the background of short-range alphas from uranium's natural radioactivity, thereby also blocking the fission fragments we would otherwise have seen. Years later I did discover the long-range alphas that are produced by fewer than 1 percent of fission events. I'm still surprised that I didn't find them at Berkeley in 1938. Had I done so, I would certainly also have quickly seen the large oscilloscope pulses due to fission fragments. I'm probably lucky to have missed the discovery of fission. I doubt if I had the maturity at twenty-seven to handle the burden of having made one of modern science's greatest discoveries. The implications of this finding disturbed Otto Hahn so profoundly, he reported later, that he seriously considered suicide.

As people arrived at the laboratory on that exciting late-January morning, we told them the news. Everyone found it hard to believe. I tracked down Robert Oppenheimer working with his entourage in his bullpen in LeConte Hall. He instantly pronounced the reaction impossible and proceeded to prove mathematically to everyone in the room that someone must have made a mistake. The next day Ken Green and I demonstrated the reaction. I invited Robert over to see the very small natural alpha-particle pulses on our oscilloscope and the tall, spiking fission pulses, twenty-five times larger. In less than fifteen minutes he not only agreed that the reaction was authentic but also speculated that in the process extra neutrons would boil off that could be used to split more uranium atoms and thereby generate power or make bombs. It was amazing to see how rapidly his mind worked, and he came to the right conclusions. His response demonstrated the scientific ethic at its best. When we proved that his previous position was untenable, he accepted the evidence with good grace, and without looking back he immediately turned to examining where the new knowledge might lead.

The extra neutrons—instantaneous "secondary" neutrons ejected in the fission process—soon became the object of a worldwide search. Frédéric Joliot, Lew Kowarski, and Hans von Halban in Paris and Fermi and Leo Szilard at Columbia University independently identified them in experiments of extraordinary difficulty. They had to find a few secondaries in the sea of neutrons that caused the uranium to fission, and they had no apparatus that could sort out fast neutrons from slow. A few years later, when Fermi had built the world's first nuclear reactor, at the University of Chicago, I amused myself by using its graphite thermal column to show the emission of secondary neutrons from uranium fission in an experiment that took half an hour from start to finish. I felt particularly stupid then because in 1939 I had a beautiful piece of apparatus—my neutron time-of-flight apparatus—that was equivalent in almost every way to a thermal column. If I had understood the importance of finding the secondaries, I could certainly have found them first.

I did decide for some reason one afternoon that I would look for them. I handled the work myself and didn't tell anyone what I was doing; but if I had continued for an hour or so I would have seen them. The boron trifluoride neutron detector in my ultracold neutron beam was set up just outside the cyclotron room on the staircase that John Lawrence's cancer patients used. I reasoned that I should look between cyclotron pulses when all fast neutrons had long ago passed by and only slow neutrons were present. If I then surrounded my counter with paraffin wax and surrounded the paraffin with a layer of cadmium, the counting rate at the detector would drop to zero. But if I put bottles of uranium compounds in the neutron beam close to the counter and the uranium gave off secondary neutrons, those particles would penetrate the cadmium, slow in the paraffin, and be detected: the counter would count them.

My reasoning was unimpeachable. I signed out several bottles of uranium oxide from the chemistry storeroom and made the necessary changes in my apparatus. Then I looked for the pulses from my counter—for about five minutes. When nothing turned up, I went back to what I had been doing a few hours before. (If only I'd known that Fermi and Joliot both had devoted all their time for the past several *months* to searching for these neutrons!) I could easily have increased the sensitivity of my experimental arrangement by a million times: by moving my counter nearer the cyclotron (× 20), by putting more uranium salt in the beam and more paraffin around the counter (× 500), by counting for an hour with and without uranium (× 100). I would have seen the secondary neutrons the same day. But I didn't; I was stupid. I didn't understand how important the experiment was, and I was too busy with several other experiments that I knew were important to see the importance for myself—I saw it only to the extent of trying a quickie experiment. Phil Abelson and I missed discovering fission, Emilio Segrè missed discovering the transuranium elements, and now I missed discovering the neutrons that accompany fission. The only consolation is that all three of these searches were pushing the state of the art.

QUESTIONS FOR DISCUSSION

1. Alvarez says that he was "affronted to see the so-called transuranium elements decaying in the wrong direction." Is the word "affronted" well chosen? What does it imply about the way scientific knowledge progresses?

2. Alvarez cites several instances of "near misses" in the discovery of fission. What might we conclude from these near misses about the process of scientific discovery? Should they be seen as failures?
3. Alvarez seems to contrast the Noddacks and Robert Oppenheimer: the former insisted "rather shabbily" on "the correctness of their work when the evidence demonstrated otherwise," whereas the latter accepted evidence contradicting his position "with good grace" and "immediately turned to examining where the new knowledge might lead," thus demonstrating "the scientific ethic at its best." Might it be argued that good scientists have in them a bit of both Oppenheimer and the Noddacks?

EVELYN FOX KELLER

A Feeling for the Organism*

Educated at Brandeis, Radcliffe, and Harvard, Evelyn Fox Keller has worked in mathematical biology and, more recently, in the history, philosophy, and psychology of science. She has taught at a number of institutions, including New York University and the University of California at Berkeley, and is currently Professor of History and Philosophy in the Program in Science, Technology, and Society at the Massachusetts Institute of Technology. Her many awards include a MacArthur Foundation Fellowship and an honorary doctorate from the University of Amsterdam.

Perhaps best known for her work on gender and science, Keller has contributed numerous articles to professional journals and is the author of several books, among them Reflections on Gender *(1985) and* Secrets of Life, Secrets of Death *(1992).*

In the following selection, taken from A Feeling for the Organism *(1983), Keller discusses the life and work of the brilliant but eccentric biologist Barbara McClintock. McClintock's work went unrecognized for much of her career, due partly to her challenging of orthodox views regarding genetic organization and partly to her idiosyncratic personal style and scientific methodology. However, her revolutionary contributions to genetic science were eventually recognized with a series of prestigious awards, culminating in a Nobel Prize (1983). In this essay, Keller focuses on McClintock's highly personal approach to scientific investigation and shows that it was an essential ingredient of her success. This piece can be fruitfully read with Root-Bernstein (this Part) since it particularizes his more general argument.*

There are two equally dangerous extremes—to shut reason out, and to let nothing else in.

PASCAL

---

*From: *A Feeling for the Organism* by Evelyn Fox Keller.

If Barbara McClintock's story illustrates the fallibility of science, it also bears witness to the underlying health of the scientific enterprise. Her eventual vindication demonstrates the capacity of science to overcome its own characteristic kinds of myopia, reminding us that its limitations do not reinforce themselves indefinitely. Their own methodology allows, even obliges, scientists to continually reencounter phenomena even their best theories cannot accommodate. Or—to look at it from the other side—however severely communication between science and nature may be impeded by the preconceptions of a particular time, some channels always remain open; and, through them, nature finds ways of reasserting itself.

But the story of McClintock's contributions to biology has another, less accessible, aspect. What is it in an individual scientist's relation to nature that facilitates the kind of seeing that eventually leads to productive discourse? What enabled McClintock to see further and deeper into the mysteries of genetics than her colleagues?

Her answer is simple. Over and over again, she tells us one must have the time to look, the patience to "hear what the material has to say to you," the openness to "let it come to you." Above all, one must have "a feeling for the organism."

One must understand "how it grows, understand its parts, understand when something is going wrong with it. [An organism] isn't just a piece of plastic, it's something that is constantly being affected by the environment, constantly showing attributes or disabilities in its growth. You have to be aware of all of that. . . . You need to know those plants well enough so that if anything changes, . . . you [can] look at the plant and right away you know what this damage you see is from—something that scraped across it or something that bit it or something that the wind did." You need to have a feeling for every individual plant.

"No two plants are exactly alike. They're all different, and as a consequence, you have to know that difference," she explains. "I start with the seedling, and I don't want to leave it. I don't feel I really know the story if I don't watch the plant all the way along. So I know every plant in the field. I know them intimately, and I find it a great pleasure to know them."

This intimate knowledge, made possible by years of close association with the organism she studies, is a prerequisite for her extraordinary perspicacity. "I have learned so much about the corn plant that when I see things, I can interpret [them] right away." Both literally and figuratively, her "feeling for the organism" has extended her vision. At the same time, it has sustained her through a lifetime of lonely endeavor, unrelieved by the solace of human intimacy or even by the embrace of her profession.

Good science cannot proceed without a deep emotional investment on the part of the scientist. It is that emotional investment that provides the motivating force for the endless hours of intense, often grueling, labor. Einstein wrote: ". . . what deep longing to understand even a faint reflexion of the reason revealed in this world had to be alive in Kepler and Newton so that they could in lonely work for many years disentangle the mechanism of celestial mechanics?" But McClintock's feeling for the organism is not simply a longing to behold the "reason revealed in this world." It is a longing to embrace the world in its very being, through reason and beyond.

For McClintock, reason—at least in the conventional sense of the word—is not by itself adequate to describe the vast complexity—even mystery—of living forms. Organisms have a life and order of their own that scientists can only partially fathom. No models we invent can begin to do full justice to the prodigious capacity of organisms to devise means for guaranteeing their own survival. On the contrary, "anything you can think of you will find." In comparison with the ingenuity of nature, our scientific intelligence seems pallid.

For her, the discovery of transposition was above all a key to the complexity of genetic organization—an indicator of the subtlety with which cytoplasm, membranes, and DNA are integrated into a single structure. It is the overall organization, or orchestration, that enables the organism to meet its needs, whatever they might be, in ways that never cease to surprise us. That capacity for surprise gives McClintock immense pleasure. She recalls, for example, the early post-World War II studies of the effect of radiation on *Drosophila:* "It turned out that the flies that had been under constant radiation were more vigorous than those that were standard. Well, it was hilarious; it was absolutely against everything that had been thought about earlier. I thought it was terribly funny; I was utterly delighted. Our experience with DDT has been similar. It was thought that insects could be readily killed off with the spraying of DDT. But the insects began to thumb their noses at anything you tried to do to them."

Our surprise is a measure of our tendency to underestimate the flexibility of living organisms. The adaptability of plants tends to be especially unappreciated. "Animals can walk around, but plants have to stay still to do the same things, with ingenious mechanisms. . . . Plants are extraordinary. For instance, . . . if you pinch a leaf of a plant you set off electric pulses. You can't touch a plant without setting off an electric pulse. . . . There is no question that plants have [all] kinds of sensitivities. They do a lot of responding to their environment. They can do almost anything you can think of. But just because they sit there, anybody walking down the road considers them just a plastic area to look at, [as if] they're not really alive."

An attentive observer knows better. At any time, for any plant, one who has sufficient patience and interest can see the myriad signs of life that a casual eye misses: "In the summertime, when you walk down the road, you'll see that the tulip leaves, if it's a little warm, turn themselves around so their backs are toward the sun. You can just see where the sun hits them and where the sun doesn't hit. . . . [Actually], within the restricted areas in which they live, they move around a great deal." These organisms "are fantastically beyond our wildest expectations."

For all of us, it is need and interest above all that induce the growth of our abilities; a motivated observer develops faculties that a casual spectator may never be aware of. Over the years, a special kind of sympathetic understanding grew in McClintock, heightening her powers of discernment, until finally, the objects of her study have become subjects in their own right; they claim from her a kind of attention that most of us experience only in relation to other persons. "Organism" is for her a code word—not simply a plant or animal ("Every component of the organism is as much of an organism as every other part")—but the name of a living form, of object-as-subject. With an uncharacteristic lapse into hyperbole, she adds: "Every time I walk on grass I feel sorry because I know the grass is screaming at me."

A bit of poetic license, perhaps, but McClintock is not a poet; she is a scientist. What marks her as such is her unwavering confidence in the underlying order of living forms, her use of the apparatus of science to gain access to that order, and her commitment to bringing back her insights into the shared language of science—even if doing so might require that language to change. The irregularities or surprises molecular biologists are now uncovering in the organization and behavior of DNA are not indications of a breakdown of order, but only of the inadequacies of our models in the face of the complexity of nature's actual order. Cells, and organisms, have an organization of their own in which nothing is random.

In short, McClintock shares with all other natural scientists the credo that nature is lawful, and the dedication to the task of articulating those laws. And she shares, with at least some, the additional awareness that reason and experiment, generally claimed to be the principal means of this pursuit, do not suffice. To quote Einstein again, "... only intuition, resting on sympathetic understanding, can lead to [these laws]; ... the daily effort comes from no deliberate intention or program, but straight from the heart."

A deep reverence for nature, a capacity for union with that which is to be known—these reflect a different image of science from that of a purely rational enterprise. Yet the two images have coexisted throughout history. We are familiar with the idea that a form of mysticism—a commitment to the unity of experience, the oneness of nature, the fundamental mystery underlying the laws of nature—plays an essential role in the process of scientific discovery. Einstein called it "cosmic religiosity." In turn, the experience of creative insight reinforces these commitments, fostering a sense of the limitations of the scientific method, and an appreciation of other ways of knowing. In all of this, McClintock is no exception. What is exceptional is her forthrightness of expression—the pride she takes in holding, and voicing, attitudes that run counter to our more customary ideas about science. In her mind, what we call the scientific method cannot by itself give us "real understanding." "It gives us relationships which are useful, valid, and technically marvelous; however, they are not the truth." And it is by no means the only way of acquiring knowledge.

That there are valid ways of knowing other than those conventionally espoused by science is a conviction of long standing for McClintock. It derives from a lifetime of experiences that science tells us little about, experiences that she herself could no more set aside than she could discard the anomalous pattern on a single kernel of corn. Perhaps it is this fidelity to her own experience that allows her to be more open than most other scientists about her unconventional beliefs. Correspondingly, she is open to unorthodox views in others, whether she agrees with them or not. She recalls, for example, a lecture given in the late 1940s at Cold Spring Harbor by Dick Roberts, a physicist from the Carnegie Institution of Washington, on the subject of extrasensory perception. Although she herself was out of town at the time, when she heard about the hostile reaction of her colleagues, she was incensed: "If they were as ignorant of the subject as I was, they had no reason for complaining."

For years, she has maintained an interest in ways of learning other than those used in the West, and she made a particular effort to inform herself about the Tibetan Buddhists: "I was so startled by their method of training and by its results that I figured we were limiting ourselves by using what we call the scientific method."

Two kinds of Tibetan expertise interested her especially. One was the way the "running lamas" ran. These men were described as running for hours on end without sign of fatigue. It seemed to her exactly the same kind of effortless floating she had secretly learned as a child.

She was equally impressed by the ability that some Tibetans had developed to regulate body temperature: "We are scientists, and we know nothing basically about controlling our body temperature. [But] the Tibetans learn to live with nothing but a tiny cotton jacket. They're out there cold winters and hot summers, and when they have been through the learning process, they have to take certain tests. One of the tests is to take a wet blanket, put it over them, and dry that blanket in the coldest weather. And they dry it."

How were they able to do these things? What would one need to do to acquire this sort of "knowledge"? She began to look at related phenomena that were closer to home: "Hypnosis also had potentials that were quite extraordinary." She began to believe that not only one's temperature, but one's circulation, and many other bodily processes generally thought to be autonomous, could be brought under the influence of mind. She was convinced that the potential for mental control revealed in hypnosis experiments, and practiced by the Tibetans, was something that could be learned. "You can do it, it can be taught." And she set out to teach herself. Long before the word "biofeedback" was invented, McClintock experimented with ways to control her own temperature and blood flow, until, in time, she began to feel a sense of what it took.

But these interests were not popular. "I couldn't tell other people at the time because it was against the 'scientific method.' . . . We just hadn't touched on this kind of knowledge in our medical physiology, [and it is] very, very different from the knowledge we call the only way." What we label scientific knowledge is "lots of fun. You get lots of correlations, but you don't get the truth. . . . Things are much more marvelous than the scientific method allows us to conceive."

Our own method could tell us about some things, but not about others—for instance, she reflects, not about "the kinds of things that made it possible for me to be creative in an unknown way. *Why* do you know? Why were you so sure of something when you couldn't tell anyone else? You weren't sure in a boastful way; you were sure in what I call a completely internal way. . . . What you had to do was put it into their frame. Wherever it came in your frame, you had to work to put it into their frame. So you work with so-called scientific methods to put it into their frame *after* you know. Well, [the question is] *how* you know it. I had the idea that the Tibetans understood this *how* you know."

McClintock is not the only scientist who has looked to the East for correctives to the limitations of Western science. Her remarks on her relation to the phenomena she studies are especially reminiscent of the lessons many physicists have drawn from the discoveries of atomic physics. Erwin Schrödinger, for example, wrote: ". . . our science—Greek science—is based on objectification. . . . But I do believe that this is precisely the point where our present way of thinking does need to be amended, perhaps by a bit of blood-transfusion from Eastern thought." Niels Bohr, the "father of quantum mechanics," was even more explicit on the subject. He wrote: "For a parallel to the lesson of atomic theory . . . [we must turn] to those kinds of epistemological problems with which already thinkers like the Buddha and Lao Tzu have been confronted, when trying to harmonize our position as spectators and actors in the great drama of existence." Robert Oppenheimer held similar views: "The general notions about human understanding . . . which are illustrated by discoveries in atomic physics are not in the nature of being wholly unfamiliar, wholly unheard of, or new," he wrote. "Even in our culture they have a history, and in Buddhist and Hindu thought a more considerable and central place." Indeed, as a result of a number of popular accounts published in the last decade, the correspondences between modern physics and Eastern thought have come to seem commonplace. But among biologists, these interests are not common. McClintock is right to see them, and herself, as oddities. And here, as elsewhere, she takes pride in being different. She is proud to call herself a "mystic."

Above all, she is proud of her ability to draw on these other ways of knowing in her work as a scientist. It is that which, to her, makes the life of science such a deeply satisfying one—even, at times, ecstatic. "What is ecstasy? I don't understand ecstasy, but I enjoy it. When I have it. Rare ecstasy."

Somehow, she doesn't know how, she has always had an "exceedingly strong feeling" for the oneness of things: "Basically, everything is one. There is no way in which you draw a line between things. What we [normally] do is to make these subdivisions, but they're not real. Our educational system is full of subdivisions that are artificial, that shouldn't be there. I think maybe poets—although I don't read poetry—have some understanding of this." The ultimate descriptive task, for both artists and scientists, is to "ensoul" what one sees, to attribute to it the life one shares with it; one learns by identification.

Much has been written on this subject, but certain remarks of Phyllis Greenacre, a psychoanalyst who has devoted a lifetime to studying the dynamics of artistic creativity, come especially close to the crux of the issue that concerns us here. For Greenacre, the necessary condition for the flowering of great talent or genius is the development in the young child of what she calls a "love affair with the world." Although she believes that a special range and intensity of sensory responsiveness may be innate in the potential artist, she also thinks that, under appropriate circumstances, this special sensitivity facilitates an early relationship with nature that resembles and may in fact substitute for the intimacy of a more conventional child's personal relationships. The forms and objects of nature provide what Greenacre calls "collective alternatives," drawing the child into a "collective love affair."

Greenacre's observations are intended to describe the childhood of the young artist, but they might just as readily depict McClintock's youth. By her own account, even as a child, McClintock neither had nor felt the need of emotional intimacy in any of her personal relationships. The world of nature provided for her the "collective alternatives" of Greenacre's artists; it became the principal focus of both her intellectual and her emotional energies. From reading the text of nature, McClintock reaps the kind of understanding and fulfillment that others acquire from personal intimacy. In short, her "feeling for the organism" is the mainspring of her creativity. It both promotes and is promoted by her access to the profound connectivity of all biological forms—of the cell, of the organism, of the ecosystem.

The flip side of the coin is her conviction that, without an awareness of the oneness of things, science can give us at most only nature-in-pieces; more often it gives us only pieces of nature. In McClintock's view, too restricted a reliance on scientific methodology invariably leads us into difficulty. "We've been spoiling the environment just dreadfully and thinking we were fine, because we were using the techniques of science. Then it turns into technology, and it's slapping us back because we didn't think it through. We were making assumptions we had no right to make. From the point of view of how the whole thing actually worked, we knew how part of it worked. . . . We didn't even inquire, didn't even see how the rest was going on. All these other things were happening and we didn't see it."

She cites the tragedy of Love Canal as one example, the acidification of the Adirondacks Lakes as another. "We didn't think [things] through. . . . If you take the train up to New Haven . . . and the wind is from the southeast, you find all of the smog from New York is going right up to New Haven. . . . We're not thinking it through, just spewing it out.... Technology is fine, but the scientists and engineers only partially think through their problems. They solve certain aspects, but not the total, and as a consequence it is slapping us back in the face very hard."

Barbara McClintock belongs to a rare genre of scientist; on a short-term view of the mood and tenor of modern biological laboratories, hers is an endangered species. Recently, after a public seminar McClintock gave in the Biology Department at Harvard University, she met informally with a group of graduate and postdoctoral students. They were responsive to her exhortation that they "take the time and look," but they were also troubled. Where does one get the time to look and to think? They argued that the new technology of molecular biology is self-propelling. It doesn't leave time. There's always the next experiment, the next sequencing to do. The pace of current research seems to preclude such a contemplative stance. McClintock was sympathetic, but reminded them, as they talked, of the "hidden complexity" that continues to lurk in the most straightforward-seeming systems. She herself had been fortunate; she had worked with a slow technology, a slow organism. Even in the old days, corn had not been popular because one could never grow more than two crops a year. But after a while, she'd found that as slow as it was, two crops a year was too fast. If she was really to analyze all that there was to see, one crop was all she could handle.

There remain, of course, always a few biologists who are able to sustain the kind of "feeling for the organism" that was so productive—both scientifically and personally—for McClintock, but to some of them the difficulties of doing so seem to grow exponentially. One contemporary, who says of her own involvement in research, "If you want to really understand about a tumor, you've got to *be* a tumor," put it this way: "Everywhere in science the talk is of winners, patents, pressures, money, no money, the rat race, the lot; things that are so completely alien . . . that I no longer know whether I can be classified as a modern scientist or as an example of a beast on the way to extinction."

McClintock takes a longer view. She is confident that nature is on the side of scientists like herself. For evidence, she points to the revolution now occurring in biology. In her view, conventional science fails to illuminate not only "how" you know, but also, and equally, "what" you know. McClintock sees additional confirmation of the need to expand our conception of science in her own—and now others'—discoveries. The "molecular" revolution in biology was a triumph of the kind of science represented by classical physics. Now, the necessary next step seems to be the reincorporation of the naturalist's approach—an approach that does not press nature with leading questions but dwells patiently in the variety and complexity of organisms. The discovery of genetic lability and flexibility forces us to recognize the magnificent integration of cellular processes—kinds of integration that are "simply incredible to our old-style thinking." As she sees it, we are in the midst of a major revolution that "will reorganize the way we look at things, the way we do research." She adds, "And I can't wait. Because I think it's going to be marvelous, simply marvelous. We're going to have a completely new realization of the relationship of things to each other."

QUESTIONS FOR DISCUSSION

1. What do you understand Keller to mean when she says (in the first paragraph) that "their own methodology allows, even obliges, scientists to continually reencounter phenomena even their best theories cannot accommodate"?
2. Keller speaks of McClintock's delight in surprise. What does this delight say about McClintock as a scientist?
3. What do you understand by the term "scientific objectivity"? Is McClintock scientifically objective?
4. McClintock always had an "exceedingly strong feeling" for the oneness of nature: "Basically, everything is one," she believed. "There is no way in which you draw a line between things." In what ways is this idea of oneness important in science? Do "subdivisions" have a usefulness?
5. Near the end of the essay, the author recounts McClintock's conversation with some science graduate students who complain that they have little time "to look and to think." Could their complaint be viewed as an echo of McClintock's broader critique of the way science has generally been practiced and taught?

SAMUEL HUBBARD SCUDDER

Learning To See

Educated at Williams College and Harvard University, Samuel Scudder was one of the leading naturalists of his time. A student of the famous zoologist Luis Agassiz (the leading American opponent of Darwin), Scudder was primarily interested in the description and classification of insects both live and fossil, perhaps reflecting the influence of his mentor. Known for the astonishing detail of his descriptions, he published something like eight hundred works. Among the most important of these are Fossil Insects of North America *and* Butterflies of the Eastern United States and Canada, *the latter representing thirty years of research.*

"Learning to See" (1874) suggests (with some humor) that, though essential to scientific discovery, careful observation does not come easily. In its emphasis on the importance of an intimate familiarity with the object of study, this essay echoes Root-Bernstein on, for example, Alexander Fleming and Keller on Barbara McClintock (both this Part).

It was more than fifteen years ago that I entered the laboratory of Professor Agassiz, and told him I had enrolled my name in the Scientific School as a student of natural history. He asked me a few questions about my object in coming, my antecedents generally, the mode in which I afterwards proposed to use the knowledge I might acquire, and, finally, whether I wished to study any special branch. To the latter

I replied that, while I wished to be well grounded in all departments of zoology, I purposed to devote myself specially to insects.

"When do you wish to begin?" he asked.

"Now," I replied.

This seemed to please him, and with an energetic "Very well!" he reached from a shelf a huge jar of specimens in yellow alcohol.

"Take this fish," said he, "and look at it; we call it a haemulon; by and by I will ask what you have seen."

With that he left me, but in a moment returned with explicit instructions as to the care of the object entrusted to me.

"No man is fit to be a naturalist," said he, "who does not know how to take care of specimens."

I was to keep the fish before me in a tin tray, and occasionally moisten the surface with alcohol from the jar, always taking care to replace the stopper tightly. Those were not the days of ground-glass stoppers and elegantly shaped exhibition jars; all the old students will recall the huge neckless glass bottles with their leaky, wax-besmeared corks, half eaten by insects, and begrimed with cellar dust. Entomology was a cleaner science than ichthyology, but the example of the Professor, who had unhesitatingly plunged to the bottom of the jar to produce the fish, was infectious; and though this alcohol had a "very ancient and fishlike smell," I really dared not show any aversion within these sacred precincts, and treated the alcohol as though it were pure water. Still I was conscious of a passing feeling of disappointment, for gazing at a fish did not commend itself to an ardent entomologist. My friends at home, too, were annoyed when they discovered that no amount of eau-de-Cologne would drown the perfume which haunted me like a shadow.

In ten minutes I had seen all that could be seen in that fish, and started in search of the Professor—who had. however, left the Museum; and when I returned, after lingering over some of the odd animals stored in the upper apartment, my specimen was dry all over. I dashed the fluid over the fish as if to resuscitate the beast from a fainting-fit, and looked with anxiety for a return of the normal sloppy appearance. This little excitement over, nothing was to be done but to return to a steadfast gaze at my mute companion. Half an hour passed—an hour—another hour; the fish began to look loathsome. I turned it over and around; looked it in the face—ghastly; from behind, beneath, above, sideways, at a three-quarters' view—just as ghastly. I was in despair; at an early hour I concluded that lunch was necessary, so, with infinite relief, the fish was carefully replaced in the jar, and for an hour I was free.

On my return, I learned that Professor Agassiz had been at the Museum, but had gone, and would not return for several hours. My fellow-students were too busy to be disturbed by continued conversation. Slowly I drew forth that hideous fish, and with a feeling of desperation again looked at it. I might not use a magnifying-glass; instruments of all kinds were interdicted. My two hands, my two eyes, and the fish; it seemed a most limited field. I pushed my finger down its throat to feel how sharp the teeth were. I began to count the scales in the different rows, until I was convinced that that was nonsense. At last a happy thought struck me—I would draw the fish; and now with surprise I began to discover new features in the creature. Just then the Professor returned.

"That is right," said he; "a pencil is one of the best of eyes. I am glad to notice, too, that you keep your specimen wet and your bottle corked."

With these encouraging words, he added:

"Well, what is it like?"

He listened attentively to my brief rehearsal of the structure of parts whose names were still unknown to me: the fringed gill-arches and movable operculum; the pores of the head, fleshy lips and lidless eyes; the lateral line, the spinous fins and forked tail; the compressed and arched body. When I had finished, he waited as if expecting more, and then, with an air of disappointment:

"You have not looked very carefully; why," he continued more earnestly, "you haven't even seen one of the most conspicuous features of the animal, which is as plainly before your eyes as the fish itself; look again, look again!" and he left me to my misery.

I was piqued; I was mortified. Still more of that wretched fish! But now I set myself to my task with a will, and discovered one new thing after another, until I saw how just the Professor's criticism had been. The afternoon passed quickly; and when, toward its close, the Professor inquired:

"Do you see it yet?"

"No," I replied, "I am certain I do not, but I see how little I saw before."

"That is next best," said he, earnestly, "but I won't hear you now; put away your fish and go home; perhaps you will be ready with a better answer in the morning. I will examine you before you look at the fish."

This was disconcerting. Not only must I think of my fish all night, studying, without the object before me, what this unknown but most visible feature might be; but also, without reviewing my discoveries, I must give an exact account of them the next day. I had a bad memory; so I walked home by Charles River in a distracted state, with my two perplexities.

The cordial greeting from the Professor the next morning was reassuring; here was a man who seemed to be quite as anxious as I that I should see for myself what he saw.

"Do you perhaps mean," I asked, "that the fish has symmetrical sides with paired organs?"

His thoroughly pleased "Of course! of course!" repaid the wakeful hours of the previous night. After he had discoursed most happily and enthusiastically—as he always did—upon the importance of this point, I ventured to ask what I should do next.

"Oh, look at your fish!" he said, and left me again to my own devices. In a little more than an hour he returned, and heard my new catalogue.

"That is good, that is good!" he repeated; "but that is not all; go on"; and so for three long days he placed that fish before my eyes, forbidding me to look at anything else, or to use any artificial aid. "Look, look, look," was his repeated injunction.

This was the best entomological lesson I ever had—a lesson whose influence has extended to the details of every subsequent study; a legacy the Professor has left to me, as he has left it to many others, of inestimable value, which we could not buy, with which we cannot part.

A year afterward, some of us were amusing ourselves with chalking outlandish beasts on the Museum blackboard. We drew prancing starfishes; frogs in mortal combat; hydra-headed worms; stately crawfishes, standing on their tails, bearing aloft umbrellas; and grotesque fishes with gaping mouths and staring eyes. The Professor came in shortly after, and was as amused as any at our experiments. He looked at the fishes.

"Haemulons, every one of them," he said. "Mr. _____ drew them."

True; and to this day, if I attempt a fish, I can draw nothing but haemulons.

The fourth day, a second fish of the same group was placed beside the first, and I was bidden to point out the resemblances and differences between the two; another and another followed, until the entire family lay before me, and a whole legion of jars covered the table and surrounding shelves; the odor had become a pleasant perfume; and even now, the sight of an old, six-inch, worm-eaten cork brings fragrant memories.

The whole group of haemulons was thus brought in review; and, whether engaged upon the dissection of the internal organs, the preparation and examination of the bony framework, or the description of the various parts, Agassiz's training in the method of observing facts and their orderly arrangement was ever accompanied by the urgent exhortation not to be content with them.

"Facts are stupid things," he would say, "until brought into connection with some general law."

At the end of eight months, it was almost with reluctance that I left these friends and turned to insects; but what I had gained by this outside experience has been of greater value than years of later investigation in my favorite groups.

QUESTIONS FOR DISCUSSION

1. After Professor Agassiz has established that the author wishes to start his study of insects immediately, Agassiz replies, "Very well!" and hands him a fish. Why a fish?
2. Early in the essay, the author quotes Professor Agassiz: "No man is fit to be a naturalist who does not know how to take care of specimens." What do you take to be the implications of this statement?
3. Agassiz always seems to be absent when Scudder needs him. What do you make of this?
4. How would Root-Bernstein (this Part) react to Scudder and his friend amusing themselves at the chalkboard?

GEORGE POINAR & ROBERTA POINAR

Million-Year-Old Cells*

A nematologist and invertebrate pathologist, George Poinar received his doctorate from Cornell University and has spent most of his career at the University of California, Berkeley. His work on nematodes (a kind of worm) led to his interest in amber as a means of studying their geolog-

The Quest for Life in Amber: The Discovery of Fossil DNA. (Excerpted from pp. ix-x; 63-69), ©1994 by George and Roberta Poinar. Reprinted by permission of Addison-Wesley Publishing Company, Inc.

ical history. This interest, in turn, prompted additional investigations on insects, arachnids, snails (and even a frog) in amber. Poinar's investigations have taken him around the world and eventually resulted in the discovery that DNA can be preserved in amber for millions of years.

An electron microscopist, Roberta Poinar has conducted research in the Departments of Zoology, Nutrition, and Entomology at the University of California, Berkeley. This work has involved the study of a whole range of organisms, including viruses, bacteria, protozoa, nematodes, insects, etc. She is the author or co-author of some ninety scientific publications. Her work has been central to the study of ancient DNA, and indeed she developed many of the laboratory techniques that made this study possible.

Since an organism's DNA is its living blueprint, studying these ancient specimens tells us something not only about the organisms themselves, but also about their evolutionary development and, indeed, about the broader evolutionary process of which we too are a product. A fictionalized version of the implications of the Poinars' discovery provided the premise for the 1993 movie "Jurassic Park."

In the piece that follows, taken from their book The Quest for Life in Amber *(1994), the Poinars convey something of the excitement that accompanied their discovery of million-year-old soft tissue preserved in amber. If soft tissue, they asked, why not DNA? Subsequent investigations provided the answer: in 1992 they were able to extract "live" DNA strands from an insect about 40 million years old. In the following year, they announced the extraction of DNA strands more than 125 million years old.*

With its first-person [I/we] account of the circumstances surrounding a significant discovery, this essay goes nicely with the piece by Alvarez (this Part). Because the Poinars' work in ancient DNA has contributed to our knowledge of the evolutionary process, their account can be read along with Hoagland's discussion (this Part) of the rapid development of molecular biology in the context of Darwin's grand, unifying idea.

Millions of years ago, trees from now vanished forests produced deposits of resin that transformed into what we know today as one of nature's most beautiful gems, amber. The forces that caused the sticky resin to slowly harden remain a mystery. While still viscous, the resin acted like flypaper and tenaciously held insects, plant parts, and even small vertebrates that touched it. Eventually the partially hardened yellow clumps, complete with their assemblage of enclosed organisms, fell from the trees to the earth. There, first leaves and debris, and later soil and rock, covered them, and eventually they were buried in layers of sandstone, limestone, or even coal. All this time, the original resin was developing the hardness, density, and melting point characteristic of amber. This durable organic gemstone outlasted the parent trees, the forests they formed, and the ecosystem of which they were a part. In some places the rock layers containing fossilized resin settled beneath the sea, and, over eons, the currents slowly loosened the amber from its grave. On the sea bottom, the amber was exposed to marine creatures that would even grow on the surface of larger pieces. Eventually the loosened amber was washed up on beaches, where humans eagerly competed with each other to collect this gold from the sea.

In other regions, the amber-bearing rock layers were shoved up into mountain ranges, where release of the gem depended on the forces of erosion. Where amber is locked up in layers of the earth, humans have dug mines to locate the "veins" and follow them as far as they can.

Whether because of its color, its feel, the mystery surrounding its formation, or the fascinating insect or plant remains it contains, amber is, and has been for centuries, valued by people. Any substance that can weather the forces of time must have magical powers. Just behold the sunlight as it reverberates off the internal fracture planes of raw amber and you will understand why many people consider it mystical.

Few things rival the beauty of fossils in amber. When examining insects entombed in this gem of the sun, you can hardly be unmoved by the wonder of seeing marvelously preserved invertebrates frozen forever in their everyday tasks—worker ants carrying food, bees with outstretched wings carrying pollen, flies mating. But it is even more miraculous that under the dissecting microscope one can easily see in some specimens intact tissues, remnants of their internal organs. This alone makes amber-embedded fossils unique, because few other fossils have been found in which soft tissues have been so well preserved. It was therefore only a matter of time before the right circumstances presented themselves—a well-preserved fossil with tissue, an experienced microscopist with an electron microscope, people with a love of amber and the willingness to look—and the idea that the fine structure of these internal tissues might also be preserved would be born and ultimately pursued, especially when our previous studies had so strongly suggested it to be true.

That someone as involved in the study of amber and amber fossils as I am would share their magic with his or her family is only natural, and amber really is a family affair in our case. One day in 1980, Roberta and I were taking turns peering down the oculars of a dissecting scope at a particularly well preserved set of Baltic amber fossils. The pile glittered in the sunlight, really fossil gold. Typically, when we are sharing some very special amber fossils, we utter many of the same exclamations of delight and wonderment that you would hear at some awe-inspiring event such as an especially vivid fireworks show. When that soon-to-be-famous female fly appeared under the spotlight of the dissecting scope, we cried, "Oh my God, look at this!" There was so much tissue, so well preserved. That fly looked like a freshly embedded specimen ready for a microscopist's knife. When we looked up at each other, the same thought was written across our faces—surely this fly's cell structure must be intact too! From then on, that mycetophilid fly became the center of our research project.

The day finally arrived when Roberta could begin work on the fly. We hoped that all the bugs—excuse the pun—had been worked out of the system in the mid-1970s, when we had worked on the fossil nematodes. I had already photographed the intact specimen, and a small piece of the amber had been chipped off and sent to Dr. Curt Beck for infrared spectroscopy analysis and confirmation that our fly was indeed in Baltic amber. We had decided to break the fly into two pieces and save one half for future studies. The piece for safekeeping was stored in an airtight container in the dark. The specimen had broken through the abdomen, and Roberta started to work on her piece. The tissue was obvious as a dark strip adjacent to the cuticle. The center of the fly, however, was an open cavity, and this proved to be the first hurdle in obtaining ultrafine sections. After a frustrating day attempting to obtain any usable sections, we decided that we would have to fill the cavity with embedding plastic before we could go any further—just another delay. It was several days later that Roberta was finally able to look at our tissue in the electron microscope. When I returned to my office after a trip to the library a note was hanging on the door: "Success!" I went immediately to the microscope room and breathlessly looked down at the lighted screen, straight onto tissue that was 40 million years old. Tissue with nuclei and organelles—mitochondria, lipid droplets and ribosomes—was there before my eyes, as well as entire muscle bands with easily identifiable components such as fibrils and mitochondria. Tracheoles, the breathing apparatus of insects, had intact linings, recognizable tubercles, and even possibly remnants of the plasma membrane.

There is something almost spiritual in a discovery such as this. It certainly makes one feel humble to look on intact cells that have been around for 40 million years. Long before human beings were even considered in the evolutionary time scale, cells existed with the same structures and organelles that they have today. How insignificant and mundane we really are.

The structure of the cells seen in our amber has a different appearance than that of cells standardly prepared for electron microscopy today. This is because scientists strive to obtain what they regard as the best fixation with cell structures having clarity and definition. Some say that electron microscopy is a study of artifacts—that tissues are frozen in time with harsh chemical fixatives, and cells are viewed when caught in a millisecond of life, with their whole life cycle extrapolated from these images. There are obviously many parallels between studying cells with the electron microscope and studying insects entombed in amber. But in the case of amber, nature used its own methods of fixation and dehydration. We feel that the fossil fly best resembles modern-day tissues that have been processed by inert dehydration using ethylene glycol. Sugars, terpenes, and other compounds in the tree sap may have combined with water in the cells to dehydrate and preserve the tissue of entrapped insects—almost a mummification process. In the past, humankind was aware of the preservative qualities of tree resins. The Egyptians used resins to embalm their nobility and wealthy citizens. Resins have antibiotic qualities that destroy fungi and bacteria and retard decaying, plus they have components that preserve tissue. Myrrh, a common embalming agent, is a mixture of the resin, gum, and essential oils from the *Commiphora* plant. This was poured into the cranial, chest, abdominal, and pelvic cavities of ancient Egyptians, and the bandages that wrapped mummies were soaked in it. Resins have also been used topically on wounds as an antiseptic and in wines to prevent spoilage. Greek Retsina wines still use resin for flavoring.

Our study of the fossil fly continued over many weeks and followed the routine cycle of ups and downs associated with electron microscope research. Although some areas of the fly were well preserved—especially a strip of hypodermal tissue immediately adjacent to the undersurface of the fly's cuticle—other tissues, further into the body cavity, lacked substructure and were composed of ghostly cell outlines. Muscle cells were the best preserved, and that has been observed by researchers in other studies of ancient tissues, such as mummies or woolly mammoths.

Preliminary results of our studies were first sent off in July 1981 to the journal *IRCS Medical Science,* where they were published later that year. More conclusive results after months more of study were published in *Science* in March 1982. This paper represented a pivotal point in our research plans. If tissues could be discovered so well preserved in amber-embedded insects 40 million years old, what else could be found? If the nuclei contained chromatin, a darkly staining structural component known to include genetic material, could DNA (deoxyribonucleic acid) actually be there too, just waiting to be discovered? If there were ribosomal structures, why then not ribonucleic acid (RNA)? We wondered if insects in older amber also had well-preserved tissue remains. On March 5, 1982, we wrote and asked Frank Carpenter at Harvard if he could send us an insect fossil in Canadian amber. On May 12 he sent us a braconid wasp from 70- to 80-million-year-old Cedar Lake amber. Roberta treated the specimen in exactly the same manner as she had treated the Baltic amber fly. The Canadian amber also sectioned with the glass knives and under the electron microscope, the sections showed well-defined tissue in the wasp's abdomen. Although the few sections we examined did not appear to pass through any nuclei, we did observe distinct sections of trachea and surrounding laminated membranous structures in partially vacuolated cytoplasm. We now had evidence that amber from different ages and plant sources could preserve insect tissues, cells, and cell organelles.

Due to our work on fossil bacteria, we had unprovable evidence, but evidence nonetheless, that even life forms might lie dormant entombed in amber. What about nucleic acids? As we were asking ourselves these questions, unknown forces were at work in other places, fitting together the pieces of a complex pattern of techniques, knowledge, and expertise that would ultimately enable us to find answers to these questions.

QUESTIONS FOR DISCUSSION

1. Traditionally, the various scientific disciplines have been fairly sharply distinguished from one another. However, someone once remarked that nature itself rejects such distinctions. Where might the authors stand on this issue? Could it be argued that they themselves have established a new discipline?
2. Describe the Poinars' attitude towards their material. Is it "scientific"?
3. What precisely was the Poinars' discovery? What is its significance?
4. In what sense was the Poinars' discovery "spiritual"? Why did it make them feel "humble"?